Clean Soil and Safe Water

NATO Science for Peace and Security Series

This Series presents the results of scientific meetings supported under the NATO Programme: Science for Peace and Security (SPS).

The NATO SPS Programme supports meetings in the following Key Priority areas: (1) Defence Against Terrorism; (2) Countering other Threats to Security and (3) NATO, Partner and Mediterranean Dialogue Country Priorities. The types of meeting supported are generally "Advanced Study Institutes" and "Advanced Research Workshops". The NATO SPS Series collects together the results of these meetings. The meetings are co-organized by scientists from NATO countries and scientists from NATO's "Partner" or "Mediterranean Dialogue" countries. The observations and recommendations made at the meetings, as well as the contents of the volumes in the Series, reflect those of participants and contributors only; they should not necessarily be regarded as reflecting NATO views or policy.

Advanced Study Institutes (ASI) are high-level tutorial courses to convey the latest developments in a subject to an advanced-level audience

Advanced Research Workshops (ARW) are expert meetings where an intense but informal exchange of views at the frontiers of a subject aims at identifying directions for future action

Following a transformation of the programme in 2006 the Series has been re-named and re-organised. Recent volumes on topics not related to security, which result from meetings supported under the programme earlier, may be found in the NATO Science Series.

The Series is published by IOS Press, Amsterdam, and Springer, Dordrecht, in conjunction with the NATO Emerging Security Challenges Division.

Sub-Series

A.	Chemistry and Biology	Springer
B.	Physics and Biophysics	Springer
C.	Environmental Security	Springer
D.	Information and Communication Security	IOS Press
E.	Human and Societal Dynamics	IOS Press

http://www.nato.int/science
http://www.springer.com
http://www.iospress.nl

Series C: Environmental Security

Clean Soil and Safe Water

edited by

Francesca F. Quercia
Institute for Environmental Protection and Research
Rome, Italy

and

Dragana Vidojevic
Serbian Environmental Protection Agency
Belgrade, Serbia

 Springer

Published in Cooperation with NATO Emerging Security Challenges Division

Proceedings of the NATO Advanced Research Workshop
on Drinking WaterProtection by Integrated Management
of Contaminated Land Belgrade, Serbia
21–23 March 2011

Library of Congress Control Number: 2011939781

ISBN 978-94-007-2242-2 (PB)
ISBN 978-94-007-2239-2 (HB)
ISBN 978-94-007-2240-8 (e-book)
DOI 10.1007/ 978-94-007-2240-8

Published by Springer,
P.O. Box 17, 3300 AA Dordrecht, The Netherlands.

www.springer.com

Printed on acid-free paper

Preface

Of actual relevance to decision makers, industry and environmental managers is the link of contaminated land management programs to the protection of drinking water resources and to potential effects of climate changes with respect to availability of these same resources.

Environmental protection has recently embraced holistic concepts of preventing threats to soil and groundwater contamination as a first priority. Methods for preventing pollution (by process improvements, monitoring and by land use and planning initiatives) coupled with integrated soil and groundwater/drinking water management at contaminated lands, represent an actual priority for existing and newly industrialized countries. This integrated approach can positively affect land, groundwater and drinking water resources protection and restoration.

At the same time Environmental security is an increasing issue in world affairs. Among the most important environmental security threats over the next 10 years, water management and environmental pollution including soil, ground water/drinking water contamination are identified. The relevance of such environmental threats will increase because of expected climate change implications.

Soil and groundwater are important components of the environment, performing many functions vital to human activities and ecosystems survival. They are viewed today as essential resources to be managed according to sustainable development principles. Groundwater pollution is closely related to soil pollution. Major causes of soil and groundwater contamination are associated with improper management of waste and leakage of hazardous substances and wastewater from factories, sewage plants and business facilities. Obsolete or abandoned factories and industrial plants and pesticide storage facilities are major sources of soil and groundwater contamination.

Groundwater represents the major drinking water supply in most countries and contamination may render groundwater unsuitable for drinking use for many years. A large number of sites affected by soil and water contamination have been identified in industrialized countries in the last 20 years, but identification is not completed yet. It is expected that in the next years more contaminated aquifers will be discovered, new contaminants will be identified and more contaminated groundwater will be discharged into wetlands, streams and lakes. Recent studies have indicated that pollution of surface water and sediments pollution in alluvial plains and coastal areas are serious problems as soil and groundwater contamination problems.

A significant part of soil and groundwater contamination in densely populated areas is associated with the operation or the abandoning of small or large obsolete factories. This happens where old or abandoned industrial plants are presently located within or nearby residential or urban areas. But it happens also where improper disposal of wastes and of obsolete pesticides, outside urban districts, affect agricultural land, residential areas and groundwater. In these cases the risk of direct exposure of local population to contamination might be significant.

In certain areas degradation might be enhanced because of climate changes which might result in lower quality and quantity of these environmental resources. Special problems are faced in particular environmental and hydrogeological settings such as arid and karst areas. So, the risk in groundwater and drinking water supplies is estimated to be higher in the following years, starting from environmentally vulnerable regions and from regions with poor water management traditions.

The need for integrated and multidisciplinary views and exchange of scientific and technical experiences in the fields of prevention, assessment and remediation aimed at the protection of drinking water from land contamination has recently been recognized.

This consideration has led to the organization of the Advanced Research Workshop entitled "Drinking Water Protection by Integrated Management of Contaminated Land" which was held in Belgrade, Serbia, in March 2011, under the NATO Science for Peace and Security Program.

The book, which collects a selection of the contributes presented at the Workshop, includes a first part on water geochemistry and groundwater vulnerability assessment, contamination and climate change impacts and contamination prevention tools. The second part provides an overview on the state of the art of soil and groundwater remediation techniques with an insight into results from a number of recent research projects. Finally, specific chapters are dedicated to national programs and progress achieved in the management of contaminated land and water safety as well as in the development of cleanup technologies and research needs.

Acknowledgements

This book has been completed after the NATO Advanced Research Workshop "Drinking Water Protection by Integrated Management of Contaminated Land" which could take place thanks to consideration and funding awarded by the NATO Science Committee.

The Editors acknowledge the NATO Science for Peace and Security Section and the NATO Publishing Unit for having provided the financial support and the technical advises needed in order to collect articles from individual authors and to edit the text.

The Editors are grateful to Walter Kovalick, US EPA, and Harald Kasamas, Environment Ministry of Austria, who, as members of the ARW Organizing Committee, promoted and facilitated this new NATO initiative as a continuation of the fruitful and beneficial experience gained in past NATO CCMS Pilot Studies on contaminated land and groundwater cleanup technologies.

Acknowledgements

Contents

Contributors

Fathi Aloui Laboratoire des Bioprocédés Environnementaux, Pôle d'Excellence Régional AUF-LBPE, Centre de Biotechnologie de Sfax, Université de Sfax, BP 1117, 3018 Sfax, Tunisia

Zaur Berishvili Material Research Institute, Faculty of Exact and Natural Sciences Iv. Javakhishvili Tbilisi State University, 3, Av. Chavchavadze, Tbilisi, 0128, Georgia

Manfred Birke Federal Institute for Geosciences and Natural Resources, BGR, Hannover, Germany, manfred.birke@bgr.de

The Eurogeosurveys Geochemistry EGG Team EuroGeoSurveys, Brussels, Belgium

Poul L. Bjerg Department of Environmental Engineering, Technical University of Denmark, Anker Engelunds Vej 1, 2800 Kgs, Lyngby, Denmark

Asher Brenner Unit of Environmental Engineering, Faculty of Engineering Sciences, Ben-Gurion University of the Negev, Be'er-Sheva 84105, Israel, brenner@bgu.ac.il

Mette M. Broholm Department of Environmental Engineering, Technical University of Denmark, Anker Engelunds Vej 1, 2800 Kgs, Lyngby, Denmark

Emanuela Bruno Water Research Institute, National Research Council of Italy (CNR-IRSA), UOS Bari Via F. De Blasio 5, Bari 70132, Italy

Mohamed Chamkha Laboratoire des Bioprocédés Environnementaux, Pôle d'Excellence Régional AUF-LBPE, Centre de Biotechnologie de Sfax, Université de Sfax, BP 1117, 3018 Sfax, Tunisia

Michelle Crimi Institute for a Sustainable Environment, Clarkson University, 8 Clarkson Avenue, Potsdam, NY, 13676-1402, USA

Alecos Demetriades Institute of Geology and Mineral Exploration, Hellas, Athens, Greece, ademetriades@igme.gr

Jacek Długosz Institute for Ecology of Industrial Areas, 6 Kossutha Street, Katowice 40-844, Poland

Gernot Döberl Contaminated Sites Department, Environment Agency, Spittelauer Lände 5, A-1090, Vienna, Austria

Joerg Drangmeister GICON – Großmann Ingenieur Consult GmbH, Tiergartenstraße 48, 01219 Dresden, Germany

Alper Elçi Department of Environmental Engineering, Dokuz Eylül University, Tinaztepe Campus, 35160 Buca-Izmir, Turkey, alper.elci@deu.edu.tr

Joerg Frauenstein Section: Soil Protection Measures, Federal Environment Agency, Woerlitzer Platz 1, Dessau-Rosslau 06844, Germany, joerg.frauenstein@uba.de

Olga Gapurova Institute of Nuclear Physics, Ulugbek, 100214 Tashkent, Uzbekistan

Boutheina Gargouri Laboratoire des Bioprocédés Environnementaux, Pôle d'Excellence Régional AUF-LBPE, Centre de Biotechnologie de Sfax, Université de Sfax, BP 1117, 3018 Sfax, Tunisia

Jochen Grossmann Section: Soil Protection Measures, Federal Environment Agency, Woerlitzer Platz 1, Dessau-Rosslau 06844, Germany

Maurizio Guerra Soil Protection Department, Institute for Environmental Protection and Research (ISPRA), Via Vitaliano Brancati 48, Rome, Italy

Tissa H. Illangasekare Environmental Science and Engineering, Colorado School of Mines, 112 Coolbaugh Hall, Golden, CO 80401, USA

Teimuraz V. Jakhutashvili Material Research Institute, Faculty of Exact and Natural Sciences, Iv. Javakhishvili Tbilisi State University, 3, Av. Chavchavadze, Tbilisi 0128, Georgia

Kestutis Kadunas Geological Survey of Lithuania, S. Konarskio str. 35, Vilnius, LT 03123, Lithuania, kestutis.kadunas@lgt.lt

Gia Kajaia Material Research Institute, Faculty of Exact and Natural Sciences, Iv. Javakhishvili Tbilisi State University, 3, Av. Chavchavadze, Tbilisi, 0128, Georgia

Fatma Karray Laboratoire des Bioprocédés Environnementaux, Pôle d'Excellence Régional AUF-LBPE, Centre de Biotechnologie de Sfax, Université de Sfax, BP 1117, 3018 Sfax, Tunisia

Harald Kasamas Division of Contaminated Land Management, Ministry of Environment, Stubenbastei, A-1010 Vienna, Austria, harald.kasamas@lebensministerium.at

Nodar P. Kekelidze Material Research Institute, Faculty of Exact and Natural Sciences, Iv. Javakhishvili Tbilisi State University, 3, Av. Chavchavadze, Tbilisi 0128, Georgia, n.kekelidze@tsu.ge

Rashid A. Khaydarov Institute of Nuclear Physics, Ulugbek, 100214 Tashkent, Uzbekistan, renat2@gmail.com

Renat R. Khaydarov Institute of Nuclear Physics, Ulugbek, 100214 Tashkent, Uzbekistan

Nana V. Khikhadze Material Research Institute, Faculty of Exact and Natural Sciences, Iv. Javakhishvili Tbilisi State University, 3, Av. Chavchavadze, Tbilisi 0128, Georgia

Mihail Kochubovski Institute of Public Health of the Former Yugoslav Republic of Macedonia, St.50 Divizija No.6, 1000 Skopje, The Former Yugoslav Republic of Macedonia, kocubov58@yahoo.com

Marek Korcz Institute for Ecology of Industrial Areas, 6 Kossutha Street, Katowice 40-844, Poland

Walter W. Kovalick, Jr. U.S. Environmental Protection Agency, 77 W Jackson Blvd. MJ-9, Chicago, IL, USA, kovalick.walter@epa.gov

Janusz Krupanek Instytut Ekologi Terenów Uprzemysłowionych, Katowice, Poland, krupanek@ietu.katowice.pl

Linda Fiedler U.S. Environmental Protection Agency, Office of Solid Waste and Emergency Response, Technology Innovation Office, Washington, DC 20460, USA, fiedler.linda@epa.gov

Jirina Machackova AECOM CZ Ltd., Trojská 92, 171 00, Prague 7, Czech Republic

Radek Malish Institute of Tropics and Subtropics, Prague, Czech Republic

Milena Zlokolica Mandić Geological Institute of Serbia, Rovinjska 12, Belgrade 11000, Serbia

John E. McCray Environmental Science and Engineering, Colorado School of Mines, 112 Coolbaugh Hall, Golden, CO 80401, USA

Najla Mhiri Laboratoire des Bioprocédés Environnementaux, Pôle d'Excellence Régional AUF-LBPE, Centre de Biotechnologie de Sfax, Université de Sfax, BP 1117, 3018 Sfax, Tunisia

Sami Mnif Laboratoire des Bioprocédés Environnementaux, Pôle d'Excellence Régional AUF-LBPE, Centre de Biotechnologie de Sfax, Université de Sfax, BP 1117, 3018 Sfax, Tunisia

Lela A. Mtsariashvili Material Research Institute, Faculty of Exact and Natural Sciences, Iv. Javakhishvili Tbilisi State University, 3, Av. Chavchavadze, Tbilisi 0128, Georgia, mtsariashvili2005@yahoo.com

Gianfranco Mulas Municipality of Portoscuso, Via Marco Polo 1, Portoscuso, CI 09010, Italy

Dietmar Müller Contaminated Sites Department, Environment Agency, Spittelauer Lände 5, A-1090, Vienna, Austria

Marine A. Nalbandyan The Center for Ecological-Noosphere Studies, The National Academy of Sciences, Abovyan str., 68, Yerevan, 0025, Armenia, marinen3@yahoo.com; ecocentr@sci.am

Petar Papić Faculty of Mining and Geology, University of Belgrade, Ðušina 7, Belgrade, Serbia

Tanja Petrović Geological Institute of Serbia, Rovinjska 12, Belgrade 11000, Serbia, tanjapetrovic.hg@gmail.com

Ivan Portoghese Water Research Institute, National Research Council of Italy (CNR-IRSA), UOS Bari Via F. De Blasio 5, Bari 70132, Italy

Clemens Reimann Geological Survey of Norway, 7491 Trondheim, Norway, clemens.reimann@ngu.no

Sami Sayadi Laboratoire des Bioprocédés Environnementaux, Pôle d'Excellence Régional AUF-LBPE, Centre de Biotechnologie de Sfax, Université de Sfax, BP 1117, 3018 Sfax, Tunisia, sami.sayadi@cbs.rnrt.tn

Robert L. Siegrist Environmental Science and Engineering, Colorado School of Mines, 112 Coolbaugh Hall, Golden, CO, 80401, USA, siegrist@mines.edu

Robert J. Steffan Shaw Environmental, Inc, 17 Princess Rd, Lawrenceville, NJ 08648, USA, rob.steffan@shawgrp.com

Jana Stojković Faculty of Mining and Geology, University of Belgrade, Ðušina 7, Belgrade, Serbia

Guy W. Sewell East Central University, 14th St, 1100 E, P-MB S-78 Ada, OK 74820, USA

Eremia V. Tulashvili Material Research Institute, Faculty of Exact and Natural Sciences, Iv. Javakhishvili Tbilisi State University, 3, Av. Chavchavadze, Tbilisi 0128

Antonella Vecchio Soil Protection Department, Institute for Environmental Protection and Research (ISPRA), Via Vitaliano Brancati 48, Rome, Italy, antonella.vecchio@isprambiente.it

Nebojša Veljković Serbian Environmental Protection Agency, Ministry of Environment Protection, Ruže Jovanovića 27a, Belgrade 11160, Serbia

Dragana Vidojević Environmental Protection Agency, Ministry of Environment, Mining and Spatial Planning, Ruže Jovanovića 27a, Belgrade 11160, Serbia, dragana.vidojevic@sepa.gov.rs

Kvetoslav Vlk Ministry of Finance of the Czech Republic, Letenska 15, Prague, Czech Republic

Faculty of Environmental Sciences, Czech University of Life Sciences, Praha 6, Suchdol, Czech Republic, kvetoslav.vlk@mfcr.cz

Michele Vurro Water Research Institute, National Research Council of Italy (CNR-IRSA), UOS Bari Via F. De Blasio 5, Bari 70132, Italy, michele.vurro@ba. irsa.cnr.it

Eleonora Wcisło Institute for Ecology of Industrial Areas, 6 Kossutha Street, Katowice 40-844, Poland, wci@ietu.katowice.pl

Zdena Wittlingerova Faculty of Environmental Science, Czech University of Life Sciences, Praha 6, Suchdol, Czech Republic

Jaroslav Zima Ministry of Finance of the Czech Republic, Letenska 15, Prague, Czech Republic

Faculty of Environmental Science, Czech University of Life Sciences, Praha 6, Suchdol, Czech Republic

Part I
Assessment and Climate Change

Chapter 1
Limitations and Challenges of Wastewater Reuse in Israel

Asher Brenner

Abstract Israel is a water-scarce country situated in a sensitive hydrological area. This has mandated a careful water resources management that integrates water resource augmentation and pollution control. Desalination of seawater and brackish groundwater, together with reclamation and reuse of municipal wastewater, has become a vital component of this concept. It is planned that by 2020, practically all municipal wastewater will be reused, mainly for agricultural irrigation. In this regard, water quality problems related to the presence of emerging trace substances, such as endocrine disrupting compounds (EDCs), may require a quaternary treatment stage that combines activated carbon adsorption, advanced oxidation processes, and desalination. The need for effluent desalination may also be required, due to salination of soil and groundwater caused by long-term irrigation with reclaimed wastewater. Since by 2020, almost 80% of Israel's fresh water supplies to the urban sector will consist of desalinated water, it will change considerably the composition of the water in use in general and consequently, that of the resulting wastewater.

1.1 Introduction

In Israel, water shortage on the one hand and the concern for water resources quality on the other, have led to the awareness that a national wastewater reclamation program must be developed. The national policy has promoted and enforced water conservation in the domestic, industrial, and agricultural sectors. However, there is a growing need for production of new water sources by desalination of seawater and

A. Brenner (✉)
Unit of Environmental Engineering, Faculty of Engineering Sciences,
Ben-Gurion University of the Negev, Be'er-Sheva 84105, Israel
e-mail: brenner@bgu.ac.il

F.F. Quercia and D. Vidojevic (eds.), *Clean Soil and Safe Water*, NATO Science for Peace and Security Series C: Environmental Security, DOI 10.1007/978-94-007-2240-8_1,
© Springer Science+Business Media B.V. 2012

brackish groundwater, as well as by reclamation and reuse of a greater proportion of municipal wastewater. Recent legislations issued by the Ministry of the Environment and by the National Water Authority require stringent quality standards for treated wastewater destined for agricultural irrigation or for disposal into rivers. These include the requirements to apply processes for nitrogen and phosphorus removal, and tertiary filtration.

Wastewater reclamation can contribute to substitution of higher quality water supplies for various applications, while preventing water pollution and health hazards. Although pollution prevention and protection of human health are global necessities in any sustainable development, the use of reclaimed wastewater for the augmentation of natural water supplies is a management issue peculiar to arid and semi-arid countries, due to increasing water shortages. If treated properly, municipal wastewater can provide valuable products, instead of being a problematic waste to be disposed and that can potentially jeopardize sensitive environments. There are many potential uses for reclaimed wastewater, some of which require high quality effluents; these include agricultural crop irrigation, recreational use, aquaculture, and groundwater recharge. In order to enable flexibility in reuse applications and to prevent pollution of surface- and ground- water, at times when there is no need for the reclaimed water, sophisticated water management should be implemented.

1.2   Water Balance in Israel

Two aquifers in Israel are the main sources of fresh water, the coastal and the mountain aquifers. Their annual production potential is approximately 300 and 350 Mm^3/Y, respectively. Other small local aquifers can add another 250 Mm^3/Y. The Sea of Galilee is a surface water source that can supply approximately 300 Mm^3/Y. There are also various local small aquifers of brackish water, especially in the southern part of Israel (The Negev Desert). This water is partly used in agriculture and industry, and its maximum production potential is approximately 300 Mm^3/Y. Most of the water is destined for the agricultural sector, which is gradually converting to the use of marginal water, especially treated effluents.

For several reasons, water management is not as simple. In Israel, as in many dry regions, most of the precipitation occurs during a short season of 4–5 months. Furthermore, there is a steep precipitation gradient from north (600–800 mm annual rainfall) to south (less than 100 mm annual rainfall) along a distance of approximately 500 km. This situation requires careful design of water conduits (from north to south) and storage reservoirs (from winter to summer). There is also uneven distribution of population (consuming water and consequently producing wastewater). The coastal plain is densely populated while the southern Negev Desert is much less so, but has the highest reserves of land for agriculture. Therefore, wastewater conveyance systems (from center to south) are required. Storage systems are also necessary for reclaimed wastewater, because it is continuously produced during the entire year, while agricultural demand is highest during the summer. Storage can be provided in open reservoirs (the most common practice in Israel) or by aquifer recharge. Both strategies affect water quality, due to chemical and biological processes occurring during long storage periods.

1.3 Future Management

Basic data regarding water demand forecast for 2020 in Israel are given in Table 1.1. The 2020 forecast of the specific municipal water consumption is 105 m^3-capita/Y, based on a population forecast of 8.5 M. In order to fulfill existing and projected water shortages, several seawater desalination plants have been designed and are gradually being erected and implemented along the Mediterranean Sea shore. Thus, by the year 2020, according to the national design, approximately 40% of the total fresh water supplies and 80% of the urban water supply will consist of desalinated water.

This change of raw water supplies will dramatically alter the mineral composition of tap water and may also affect the composition of the reclaimed wastewater.

According to the Israeli strategy, by the year 2020 (and even before), approximately 50% of agricultural water demand will be provided by reclaimed wastewater, assuming that agriculture maintains its present size. This means that all municipal wastewater should be treated and reused, mainly in agriculture, but also for recreational purposes and for river restoration. Since 60–70% of the municipal water consumption results in sewage that can be collected, treated and reused, this is a very reliable water source, since domestic water supply will always be of high priority. According to a recent survey conducted by the Israel Water Authority (2008), the total amount of wastewater in Israel for that year was approximately 500 Mm^3/Y, 88% of which was municipal wastewater. Seventy-five percent of the municipal wastewater was reused in that year. The policy to reuse increasing amounts of marginal water (including effluents, brackish water and storm-water) in place of fresh water is illustrated in Fig. 1.1.

Another major issue that has to be considered in extensive reuse of reclaimed wastewater for agricultural irrigation is the salination of soil and groundwater.

Table 1.1 Year 2020 water demand forecast in Israel (Mm^3/Y)

	Fresh natural	Desalinated	Effluents	Brackish and runoff	Total	% of grand total	% of fresh[a]
Agriculture	450	/	550	100	1,100	43%	25%
Urban	200	700	/	/	900	35%	50%
Industry	100	/	/	50	150	6%	6%
Others[b]	350	/	50	/	400	16%	19%
Total	1,100	700	600	150	2,550	100%	
% of grand total	43%	27%	24%	6%	100%		

[a] % of fresh water consumption of the total fresh water consumption including desalinated water (1,800 Mm^3/Y)
[b] Agreements with state neighbors, aquifer rehabilitation, and nature conservation

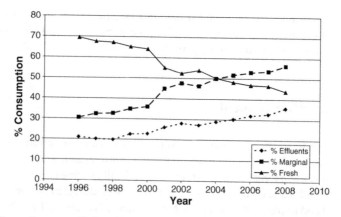

Fig. 1.1 Types of water used in Israel for agriculture (Data obtained from the Israel Water Authority)

The dilution of natural water supplies by a high fraction of desalinated water will reduce salt concentration, but will not reduce salt quantity accumulated in irrigated soils.

Under effluent irrigated agriculture, a certain amount of excess water is required to percolate through the root zone, to remove the salts that have accumulated as a result of evapotranspiration. This requires excess water for irrigation. However, this might transfer the problem of salts accumulation to groundwater down below. Thus, desalination of effluent may become necessary in order to prevent soil and groundwater salination, and to conserve excess irrigation water needed to leach salts out of the root zone.

Furthermore, the comprehensive reuse of treated wastewater may ultimately cause a long-term buildup of toxic chemicals in the closed-loop cycle of water supply and wastewater treatment and reuse [2, 4]. The problems of emerging trace organic constituents, such as pharmaceuticals and personal care products (PPCPs), some of which are considered endocrine disrupting compounds (EDCs), may require the use of a quaternary treatment stage as an integral part of reclamation schemes. In this regard, the era of BOD, as a control parameter for effluent organic content, has actually ended. In addition, advanced monitoring tools and methods are required to identify the fate of a variety of specific substances at the μg/L and even ng/L levels along treatment trains. Effluent desalination should thus be combined in the quaternary treatment stage together with carbon adsorption and/or advanced oxidation processes, in order to effectively remove problematic trace substances. It has been shown that desalination alone, by nanofiltration or by reverse osmosis, is not capable of absolute removal of a variety of trace compounds [3]. This concept is illustrated in Fig. 1.2.

Future management of multi-quality water resources (including natural water supplies, desalinated water and reclaimed wastewater) also requires some modification to the traditional upstream treatment strategies commonly applied for drinking water supplies. For example, in desalinated water, the levels of alkalinity and essential minerals, such as calcium and magnesium, are very low. Therefore, desalinated

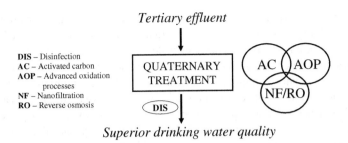

DIS – Disinfection
AC – Activated carbon
AOP – Advanced oxidation
        processes
NF – Nanofiltration
RO – Reverse osmosis

Tertiary effluent

QUATERNARY
TREATMENT

AC   AOP

NF/RO

DIS

Superior drinking water quality

Fig. 1.2 Conceptual approach of quaternary treatment

water may be associated with inferior taste and corrosion problems that result in the release of metal colloids into water distribution pipes. In addition, the water-intake of these essential nutrients will be reduced dramatically in some populations.

Water treatment processes can affect mineral concentrations and contribute to the total intake of calcium and magnesium for some individuals. Therefore, the Israeli Ministry of Health has proposed new quality standards for desalinated water, requiring the application of post-treatment for the conditioning of desalinated water, mainly through dissolution of calcium carbonate. This process can supply the proposed quality, including: alkalinity >80 mg/L as $CaCO_3$; $80 < Ca^{2+} < 120$ mg/L as $CaCO_3$; calcium carbonate precipitation potential (CCPP)>3 mg/L as $CaCO_3$; pH<8.5. These quality standards actually cover the need to account for nutritional supply of calcium (>50 mg/L as $CaCO_3$). The proposed water quality standards do not include a requirement for magnesium addition, on the assumption that this mineral's intake-requirements can be obtained from food products more easily than calcium. However, there have been complaints by farmers who use effluents originating from desalinated water that lack of magnesium affects plant yield and quality.

Addition of alkalinity to drinking water may also be crucial for wastewater treatment schemes employing alkalinity-consumption processes, such as nitrification. The current proposed standard for alkalinity of desalinated water (based on human health considerations, related to corrosion of metal pipes) may not be sufficient to support wastewater nitrification, since typical municipal wastewater in Israel is relatively concentrated (BOD=400 mg/L, TKN=70 mg/L). Alkalinity is usually increased during municipal water use; however further measures may be required to remedy the lack.

There is another problem related to the quality of desalinated water, based on its conversion after primary use to municipal wastewater destined for reclamation and reuse. Boron toxicity to plants may limit the application of reclaimed wastewater originating from seawater, because of the high content of boron (approximately 5 mg/L) and its limited rejection in conventional reverse osmosis processes [1]. The new desalination plants planned in Israel are therefore required by the Israel Water Authority to upgrade their processes to reduce boron levels to 0.3 mg/L. In addition, source control measures have been applied by the Israeli Ministry of Environmental Protection, enforcing a gradual reduction of boron content of detergents for washing machines that constitute a major source of boron in municipal wastewater.

Another source control policy enforced in Israel (by legislation) to reduce the increase in salinity of municipal wastewater (a more than two fold increase has been recorded in several municipal wastewater systems) is to forbid the disposal of industrial brines and ion exchange regeneration liquids into municipal sewers [5]. This has partially prevented point-source intrusions of salts into municipal wastewater and, consequently, has reduced the total salinity of many waste streams. In addition, the legislation has led many plants to convert to reverse osmosis in place of ion exchange, to reduce the quantities of salts requiring disposal. These measures, together with more stringent source control of heavy metals, have improved the overall quality of both effluent and sludge, but cannot be considered a solution of the problem sufficient to replace advanced wastewater treatment.

1.4   Conclusions

Future management of multi-quality water resources in Israel is indeed a complex issue, incorporating several measures such as modification of traditional water treatment schemes and quality standards, upgrading of wastewater treatment processes, and source control. Specific conclusions can be summarized as follows:

- Wastewater reclamation and reuse in Israel are a national policy that can solve three problems:
  - mitigate (partly) the shortage of water;
  - prevent pollution of natural water resources;
  - preserve agriculture in its current size.
- Long-term utilization of effluents for agricultural irrigation may cause salination of soil and groundwater. This problem may require effluent desalination.
- Emerging substances accumulating in the closed-loop water cycle (use-reuse) may require the use of a quaternary treatment stage.
- The quaternary stage should include a combination of advanced processes (such as NF, RO, AOP, AC) to result in a superior water quality. The technological formula should further be investigated.
- Water treatment alternatives and quality standards should be based also on reuse considerations.
- Source control of municipal wastewater "production" is an important means to enable a sustainable reuse.

References

1. Hyung H, Kim J-H (2006) A mechanistic study on boron rejection by sea water reverse osmosis membranes. J Membr Sci 286:269–278
2. Kolpin DW, Furlong ET, Meyer MT, Thurman EM, Zaugg SD, Barber LB, Buxton HB (2001) Pharmaceuticals, hormones, and other organic wastewater substances in U.S. streams, 1999–2000: a national reconnaissance. Environ Sci Technol 36(6):1202–1211

3. Sahar E, David I, Gelman Y, Cikurel H, Aharoni A, Messalem R, Brenner A (2011) The use of RO to remove emerging micropollutants following CAS/UF or MBR treatment of municipal wastewater. Desalination 273:142–147
4. Ternes TA, Joss A (2006) Human pharmaceuticals, hormones and fragrances: the challenge of micropollutants in urban water management. IWA, London
5. Weber B, Juanico M (2004) Salt reduction in municipal sewage allocated for reuse: the outcome of a new policy in Israel. Water Sci Technol 50(2):17–22

Chapter 2
Research of Physical-Chemical Parameters in Drinking (Tap) Water in Tbilisi City and Its Close Regions

Nodar P. Kekelidze, Teimuraz V. Jakhutashvili, Lela A. Mtsariashvili, and Nana V. Khikhadze

Abstract Research of drinking water quality, especially in a large settlement like Tbilisi (the largest center of population in Georgia) by using modern methods and equipment is the actual task of research of water resources ecological conditions. This work presents the data from research of drinking water quality parameters, the main five physical-chemical parameters – specific electrical conductivity, total dissolved solids, salinity, pH and water temperature, in different districts of Tbilisi city and its suburbs. In the course of a year we received data for the controlled points of Tbilisi city vary as follows: pH in the range of 6.75–7.51; total dissolved solids – from 115 up to 185 mg/L; specific electrical conductivity from 231 up to 369 μS/cm; salinity from 0.12 up to 0.18 ppt. The physical-chemical parameters of water quality for Tbilisi city suburb, in the course of a year vary as follows: pH in the range of 6.7–7.5; total dissolved solids from 142 up to 193 mg/L; specific electrical conductivity from 284 up to 386 μS/cm; salinity from 0.14 up to 0.19 ppt. The received data for pH and total dissolved solids for water don't exceed the values of maximum permissible concentration. Some peculiarities of the collected data are discussed in this paper.

2.1 Introduction

Water is a special mineral of the Earth, which "controls" its climate and is necessary to life. In spite of the fact that water seems to be a never-ending mineral, water basin pollution increases quickly, and humans are faced with the complex problem of clean water protection. Water changes as does our common environment.

N.P. Kekelidze (✉) • T.V. Jakhutashvili • L.A. Mtsariashvili • N.V. Khikhadze
Material Research Institute, Faculty of Exact and Natural Sciences, Iv. Javakhishvili
Tbilisi State University, 3, Av. Chavchavadze, Tbilisi 0128, Georgia
e-mail: n.kekelidze@tsu.ge; mtsariashvili2005@yahoo.com

F.F. Quercia and D. Vidojevic (eds.), *Clean Soil and Safe Water*, NATO Science for Peace and Security Series C: Environmental Security, DOI 10.1007/978-94-007-2240-8_2, © Springer Science+Business Media B.V. 2012

Table 2.1 Normalized values of physical-chemical parameters for potable water reported by some organizations and countries

Parameter	Measuring unit	WHO	USEPA	EC	Russia	Georgia[a, b]
pH	pH value	–	6.5–8.5	6.5–8.5	6–9	6–9
Total mineralization (TDS)	mg/L	1,000	500	1,500	1,000	100–1,500
Electro-conductivity (at 20°C)	μS/cm	–	–	–	–	–
Temperature	°C	–	–	25	–	<30

[a] Normative document [7]
[b] Technical Requirements to Quality of Drinking Water [9]

Many different substances, sometimes unknown in the past and hardly detected, flow into rivers, seas and in oceans. Composition of surface waters also changes because of water runoff which transports different substances from the rural fields and agricultural holdings [3, 5].

Hence, it is necessary to create and introduce new methods for analyzing water pollutants, and to plan a regular study of the different water quality parameters.

Characteristics of concern are water quality physical-chemical parameters, such as pH and the total mineralization (or Total Dissolved Solids), as well as temperature, electro-conductivity and salinity.

Water pH is one of the major indicators of water quality, defining the character of the chemical and biological processes occurring in water.

The level of total water mineralization is in many respects determined by the character of surrounding soil-geological structures which essentially varies in different geographical regions and basically defines the electro-conductivity of natural waters. Because of this relationship water mineralization can be estimated with a certain accuracy from the value of its electro-conductivity.

Water temperature is the major factor influencing the current physical, chemical, biochemical and biological processes in water. By WHO (World Health Organization) recommendations water temperature "should be acceptable" given that, in consumer's view, cold water, as a rule, is considered tastier.

In Table 2.1 normalized values of these parameters for potable water, as reported by some international organizations and countries are given. Not all parameters (for example, electro-conductivity) are normalized.

In the present work the data for some physical-chemical parameters of drinking (tap) water in various areas of Georgia, such as Tbilisi city and its suburbs, are given.

Some water quality physical-chemical parameters of the water supplies in Tbilisi city (Zhinvali and Tbilisi reservoirs) were investigated in some scientific publications [4, 8].

In the work from Ghogheliani et al. [4], eco-chemical and hygienic conditions of Zhinvali and Tbilisi reservoirs are considered. They have shown that values of water quality basic parameters (pH 7.7–7.8; total dissolved solids 222.4–255 mg/L) of Zhinvali and Tbilisi reservoirs don't exceed the values of normalized levels of maximum permissible concentrations in water supplies. Investigated water, in most cases, is classified as very clean water.

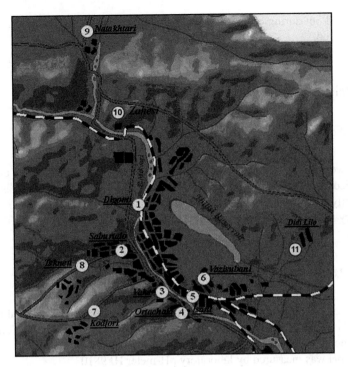

Fig. 2.1 Scheme of control points location in the territory of Tbilisi city and its suburbs

In the work from Supatashvili [8] some physical-chemical parameters of water quality in Sioni and Tbilisi reservoirs (in particular, in Tbilisi reservoir pH value is 8.25 and in Sioni reservoir pH value is 8.12) are given.

There is no water quality literature data for Tbilisi city and its suburbs water supply (tap water) system.

2.2   Research Tasks and Objectives

In Tbilisi city area drinking (tap) water samples were collected once per month (for the period of 2009) at the following sampling locations: Dighomi (1), Saburtalo (2), Vake (3), Ortachala (4), Isani (5), Vazisubani (6). Drinking (tap) water samples were also collected in the following regions surrounding Tbilisi city: settlement Kojori (7), settlement Tskneti (8), village Natakhtari (9), suburb Zahesi (10) and suburb Lilo (11) (see Fig. 2.1).

At above mentioned control points five physical-chemical parameters of water quality – water pH, temperature (T), electro-conductivity (Ec), Total Dissolved Solids (TDS) and Salinity (S) – were measured.

Tbilisi city water samples measurements were carried out once per month and 72 water samples were analyzed. Tbilisi city suburbs water samples measurements were carried out during each field survey (once per quarter) and 20 water samples were analyzed.

2.3 Brief Description of Methodology

2.3.1 Sample Collection

Samples were selected directly from the water supply system. Sampling of tap water was carried out simultaneously in different districts of Tbilisi (sampling was carried out directly in containers for measurement). Then samples were transferred to a laboratory for analytical measurements.

2.3.2 Measurements

In Tbilisi city water samples physical-chemical parameters of water quality were measured in a laboratory in the same day of sampling. In regions close to Tbilisi city measurements were carried out during field surveys.

Water pH was measured by laboratory pH-meter HI 98103.

Water temperature, electro-conductivity, total dissolved solids and salinity were measured with a Sension 5 conductometer and estimated with a calculation method which compares the data in the device memory (electro-conductivity and temperature) with measured data.

2.4 Research Results

2.4.1 Tbilisi City

Data variations at control points. Annual average (av), minimum (min) and maximum (max) values, and range of variation (difference D=max−min) values of drinking (tap) water physical-chemical parameters at selected control points are given in Table 2.2 and in Fig. 2.2, showing that:

- average value of water pH is 7.19, while both minimum (6.75) and maximum values (7.51) are found in Vake; higher values are registered at Dighomi, Isani and Vazisubani control points, while in Saburtalo, Vake and Ortachala lower values are observed. The maximum value (7.51) and the largest variation (0.76) are observed in Vake, while the smallest variation (0.51) is observed in Vazisubani;
- the estimated average value of TDS is 158 mg/L; the minimum value (115 mg/L) was observed in Dighomi and maximum (185 mg/L) in Vake; values are higher

Table 2.2 Tbilisi city drinking (tap) water physical-chemical parameters average annual (av), minimal (min) and maximal (max) values, and their changing (D) range values depending on control points

Parameter		Control points						All control points
		Dg	Sb	Vk	Or	Is	Vz	
pH	av	7.21	7.19	7.17	7.14	7.16	7.25	7.19
	min	6.91	6.83	6.75	6.85	6.81	6.99	6.75
	max	7.47	7.50	7.51	7.45	7.42	7.50	7.51
	D	0.56	0.67	0.76	0.6	0.61	0.51	0.76
T, °C	av	18.9	18.6	18.6	18.6	18.7	19.1	18.8
	min	14.7	14.8	15.2	14.3	14	15.5	14.0
	max	23.7	23.2	22.7	23.7	23.6	23.6	23.7
	D	9.0	8.4	7.5	9.4	9.6	8.1	9.70
Ec, μS/cm	av	284	326	333	323	317	313	316
	min	231	293	290	272	288	280	231
	max	339	355	369	362	347	349	369
	D	108	62	79	90	59	69	138
TDS, mg/L	av	142	163	167	161	159	157	158
	min	115	147	145	134	144	140	115
	max	170	177	185	181	174	180	185
	D	55	30	40	47	30	40	70
S, ppt	av	0.14	0.16	0.17	0.16	0.16	0.16	0.16
	min	0.12	0.15	0.15	0.14	0.14	0.14	0.12
	max	0.17	0.18	0.18	0.18	0.17	0.17	0.18
	D	0.05	0.03	0.03	0.04	0.03	0.03	0.06

at Saburtalo, Vake and Ortachala control points; lower values are observed in Dighomi, Isani and Vazisubani where an inverse dependence with respect to pH was observed; the largest range of variation (55 mg/L) is recorded in Dighomi, the smallest (30 mg/L) in Isani. Similar features were observed for electro-conductivity and salinity, as measurements techniques are interconnected.

Data variation with time. Monthly average (av), minimum (min) and maximum (max) values, and range of variation (difference D=max−min) values of physical-chemical parameters in Tbilisi city drinking (tap) water control points are given in Table 2.3 and Fig. 2.3, showing that:

- a decrease of pH values in May, August and December and an increase during other periods. The maximum value (7.51) is recorded in February, and minimum (6.75) in August; the largest range of variation (0.66) is observed in February, and the lowest (0.11) in December.
- TDS values decrease in December–January and July–August. The minimum value (115 mg/L) is observed in July, and maximum (185 mg/L) in March; the largest range of variation (52 mg/L) is observed in July, the lowest (15 mg/L) in March.

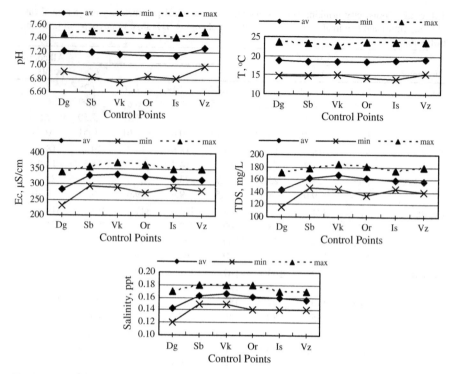

Fig. 2.2 Tbilisi city drinking (tap) water physical-chemical parameters average annual, minimum and maximum values at different control points (*Dg* Dighomi; *Sb* Saburtalo; *Vk* Vake; *Or* Ortachala; *Is* Isani; *Vz* Vazisubani). Note: Temperature data have just an informative character and were necessary for adjusting Ec and pH measurements which were carried out under laboratory conditions

2.4.2 Tbilisi Suburbs

Data variations at control points. Average annual (av), minimum (min) and maximum (max) values, and range of variation (D=max−min) for drinking (tap) water physical-chemical parameters at selected control points in Tbilisi city suburbs are given in Table 2.4 and Fig. 2.4, showing that:

- average value of water pH is 7.08, minimum (6.70) value was recorded in suburb Tskneti and maximum value (7.50) was recorded in Lilo; values tend to increase in Kojori and Lilo, and to decrease in Tskneti, Natakhtari and Zahesi. The greatest range of variation (0.50) is observed in Tskneti, Lilo and Zahesi, the lowest (0.20) in Kojori.
- average value of TDS is 158 mg/L, minimum value (142 mg/L) was recorded in Zahesi, and maximum (193 mg/L) in Natakhtari; TDS values tend to increase in Natakhtari and Zahesi, and to decrease in Tskneti, Kojori and Lilo; the largest range of variation (50 mg/L) is observed in Natakhtari, the smallest (27 mg/L) in

Table 2.3 Monthly average (av), minimum (min) and maximum (max) values, and range of variation (D) values of physical-chemical parameters in Tbilisi city drinking (tap) water versus time

Parameter		I	II	III	IV	V	VI	VII	VIII	IX	X	XI	XII
		Month											
pH	av	7.40	7.32	7.33	7.13	7.29	7.26	7.20	6.89	7.15	7.18	7.13	6.95
	min	7.31	6.85	7.25	6.95	7.05	7.05	7.13	6.75	6.93	7.13	7.03	6.90
	max	7.5	7.51	7.50	7.29	7.45	7.47	7.29	7.15	7.32	7.25	7.2	7.01
	D	0.19	0.66	0.25	0.34	0.4	0.42	0.16	0.4	0.39	0.12	0.17	0.11
T, °C	av	16.6	14.8	16.6	15.1	20	23.4	20.8	23	20.8	19.3	19	15.7
	min	15.8	14	15.9	14.3	19.5	22.7	20.1	22.7	20.1	19.1	17.8	15.6
	max	17.2	15.5	17.6	16.2	20.4	23.7	21.6	23.7	21.6	19.6	19.3	15.8
	D	1.4	1.5	1.7	1.9	0.9	1.0	1.5	1.0	1.5	0.5	1.5	0.2
Ec, μS/cm	av	333	341	350	346	335	331	294	290	289	300	297	288
	min	305	327	339	322	317	304	231	232	252	260	267	257
	max	352	364	369	360	355	362	334	329	308	320	328	302
	D	47	37	30	38	38	58	103	97	56	60	61	45
TDS, mg/L	av	166	171	175	173	170	165	147	145	144	150	149	144
	min	152	164	170	161	161	152	115	116	126	130	134	128
	max	176	182	185	180	180	181	167	165	154	160	164	151
	D	24	18	15	19	19	29	52	49	28	30	30	23
S, ppt	av	0.17	0.17	0.17	0.17	0.17	0.17	0.15	0.15	0.14	0.15	0.15	0.15
	min	0.15	0.16	0.17	0.16	0.16	0.15	0.12	0.12	0.13	0.13	0.13	0.13
	max	0.18	0.18	0.18	0.18	0.18	0.18	0.17	0.16	0.15	0.16	0.16	0.15
	D	0.03	0.02	0.01	0.02	0.02	0.03	0.05	0.04	0.02	0.03	0.03	0.02

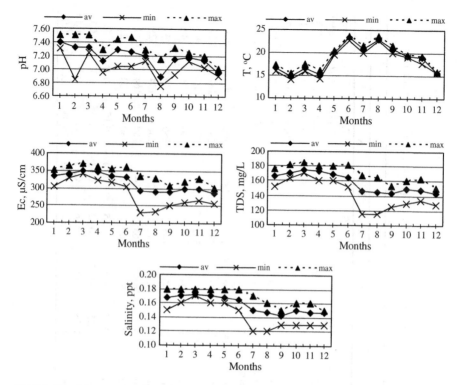

Fig. 2.3 Monthly average (*av*), minimum (*min*) and maximum (*max*) values of physical-chemical parameters in Tbilisi city drinking (tap) water versus time

Tskneti and Lilo (similar features were observed for electro-conductivity and salinity as measurement techniques are interconnected).

- average value of temperature is 16.5°C, minimum value (9.5°C) was recorded in Tskneti and maximum (22.3°C) in Lilo.

Data variation with time. Monthly average (av), minimum (min) and maximum (max) values, and range of variation (D = max−min) values of physical-chemical parameters in Tbilisi suburbs drinking (tap) water at the selected control points, are given in Table 2.5 and Fig. 2.5, showing that:

- the trend of pH values with time is to decrease in March and September and to increase in May and November; maximum value (7.50) is recorded in May, and minimum (6.70) in March; the largest range of variation (0.7) is observed in May and the smallest (0.20) in November;
- the tendency of TDS values with time is to decrease in May, September and November. The minimum value 142 mg/L is recorded in November and maximum (193 mg/L) in March; the largest range of variation (26 mg/L) is observed in March and the lowest (4 mg/L) in November;

Table 2.4 Tbilisi city suburbs drinking (tap) water physical-chemical parameters annual average (av), minimum (min) and maximum (max) values, and their range of variation (D) at selected control points

Parameter		Control points					All control points
		Tskn.	Koj.	Nat.	Lilo	Zah.	
pH	av	6.9	7.1	7.0	7.3	7.1	7.08
	min	6.7	7	6.8	7	6.9	6.70
	max	7.2	7.2	7.2	7.5	7.4	7.50
	D	0.5	0.2	0.4	0.5	0.5	0.8
T, °C	av	15	15	18.2	16.8	17.5	16.5
	min	9.5	10.1	13.3	12.6	13.6	9.50
	max	21.1	18.2	22.1	22.3	21.5	22.3
	D	11.6	8.1	8.8	9.7	7.9	12.8
Ec, μS/cm	av	315	318	322	313	316	317
	min	291	289	286	286	284	284
	max	346	353	386	340	352	386
	D	55	64	100	54	68	102
TDS, mg/L	av	158	159	161	156	158	158
	min	146	144	143	143	142	142
	max	173	177	193	170	176	193
	D	27	33	50	27	34	51
S, ppt	av	0.16	0.16	0.16	0.16	0.16	0.16
	min	0.15	0.14	0.14	0.14	0.14	0.14
	max	0.17	0.18	0.19	0.17	0.18	0.19
	D	0.02	0.04	0.05	0.03	0.04	0.05

- the temperature trend with time is to increase during the summer period (May, September) and to decrease in winter (November, March). The minimum value (9.5°C) is observed in March and maximum (22.3°C) in September; the largest range of variation (6.0°C) is observed in May and the lowest (3.9°C) in November.

2.5 Analysis

It is apparent from the results on the range of pH variability that, according to the accepted system of pH level classification [2], practically all sampled waters correspond to the group of neutral waters (6.5–7.5). According to recorded TDS values all water samples correspond to the category of ultra fresh waters (<200 mg/L).

In Table 2.6 a comparison of some literature data for natural water sources of potable water in Tbilisi city and its suburbs with the data shown in the present work is reported.

It is apparent from these data that pH, TDS and Ec values collected in the present work for drinking (tap) water are appreciably lower than similar parameters for natural reservoirs waters (belonging to alkalescent groups and fresh waters categories).

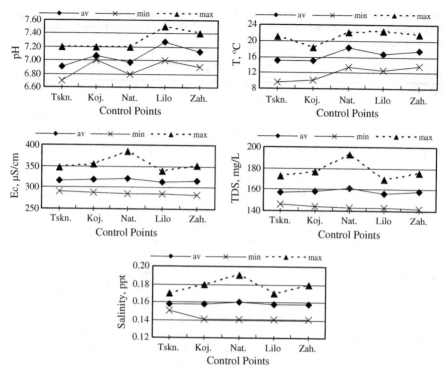

Fig. 2.4 Tbilisi city suburbs drinking (tap) water physical-chemical parameters annual average, minimum and maximum valuesat selected control points (*Tskn.* Tskneti; *Koj.* Kojori; *Nat.* Natakhtari; *Zah.* Zahesi; *Lilo* Lilo)

This circumstance can be explained by considering that reported literature data do not include mean annual estimates but rather reflect values at a given time. Appreciable difference of pH values for the same object – Tbilisi reservoir – between the data from two different publications may confirm this remark. However it is impossible to exclude that the processing of natural waters in the city's water supply system may affect these parameters.

There is also another factor influencing drinking water parameters. To the city's water supply a significant contribute is brought by groundwater from artesian wells in Natahtari area. In the city water system flow waters originating from different sources and in different areas the prevalence of waters from a specific source can be observed. This fact explains the observed larger values of pH in some areas of the city (Dighomi, Isani and Vazisubani) and lower in others (Saburtalo, Vake and Ortachala). As it has been shown in an earlier work [6], an appreciable distinction in the content of radon in water in various areas of the city was connected with the same circumstance.

In particular collected results suggest that in Saburtalo, Vake and Ortachala waters from Natakhtari source prevail while in Dighomi, Isani and Vazisubani surface waters from reservoirs prevail. Radon does not influence water quality; however

Table 2.5 Monthly average (av), minimum (min) and maximum (max) values, and range of variation (D) values of physical-chemical parameters in Tbilisi suburbs drinking (tap) water versus time

Parameter		Months			
		March	May	September	November
pH	av	7.0	7.1	7.1	7.1
	min	6.7	6.8	6.8	7.0
	max	7.2	7.5	7.4	7.2
	D	0.5	0.7	0.6	0.2
T, °C	av	11.8	18.3	21	14.8
	min	9.5	16.1	18.2	12.3
	max	13.6	22.1	22.3	16.2
	D	4.1	6.0	4.1	3.9
Ec, µS/cm	av	354	324	301	287
	min	335	316	290	284
	max	386	340	308	291
	D	51	24	18	7
TDS, mg/L	av	177	162	151	144
	min	167	158	145	142
	max	193	170	154	146
	D	26	12	9	4
S, ppt	av	0.18	0.16	0.15	0.14
	min	0.17	0.16	0.15	0.14
	max	0.19	0.17	0.15	0.15
	D	0.02	0.01	0.00	0.01

distinction in the nature of the waters incoming in the city water supply system can be reflected in other parameters causing observed pH and TDS trends and range of their changes at selected control points locations.

Observed tendencies in seasonal variations are due to the influence of atmospheric precipitation (in the form of a rain or flood waters). Their intensity increased during the periods when a pH decreasing trend (and TDS increasing) was observed. It is known that pH in atmospheric precipitation is in the range 4.6–6.1 [1]. Mixing up with groundwater and surface waters can cause the observed trends in change of pH and TDS values with time.

The results observed in suburbs' tap water are practically similar to those observed in the city. Temperature data correspond to expected tendencies in corresponding seasons (increase during the summer period and decrease in winter). Lower values recorded in Tskneti and Kojori suburbs, reflect their higher altitude above sea level that causes some decrease of ground temperature, and, consequently, of tap water temperature.

It is necessary to note, that all water pH and TDS recorded data don't exceed maximum permissible concentration values (see Table 2.1). Data on salinity do not represent special interest; this parameter is generally used for the estimation of salty waters and, according to Alekin [1], for natural waters it oscillates from 25 up to 50‰.

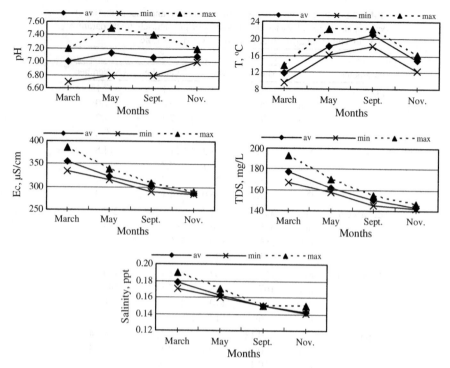

Fig. 2.5 Monthly average (*av*), minimum (*min*) and maximum (*max*) values of physical-chemical parameters in Tbilisi suburbs drinking (tap) water versus time

Table 2.6 Data of physical-chemical parameters of water quality in various water sources in Tbilisi city area

#	Control points	pH	T, °C	TDS, mg/L	Ec, µS/cm	S	References
1	Zhinvali reservoir	7.8	2.5–20.0	222.4	400	–	Ghogheliani et al. [4]
2	Tbilisi reservoir	7.7	8–20	255	400	–	Ghogheliani et al. [4]
3	Tbilisi reservoir	8.25	–	–	–	–	Supatashvili [8]
4	Sioni reservoir	8.12	–	–	–	–	Supatashvili [8]
5	Tbilisi (tap water)	7.19	18.8	158	316	0.16	In present study
6	Suburbs of Tbilisi (tap water)	7.07	16.5	158	317	0.16	In present study

2.6 Conclusions

1. In this work regular studies of some physical-chemical parameters – pH; Temperature; Electro-conductivity; Total Dissolved Solids; Salinity – in tap water of the largest center of population in Georgia –Tbilisi and its suburbs – have been carried out;

2. the following range of values have been estimated: pH- in the range 6.70–7.51; Temperature – from 9.5°C up to 23.7°C; Electro-conductivity – from 231 up to 386 μS/cm; Total Dissolved Solids from 115 up to 193 mg/L; Salinity from 0.12 up to 0.19 ppt . These values correspond to values established for drinking water in the legislation. Drinking water in all control points belongs to the neutral group and ultra fresh category;
3. Some trends and parameters variability, versus time and versus control points location, have been observed and discussed.

Acknowledgment Present work has been completed with the support of Georgia National Science Foundation (Grant # GNSF/PRES08/5-331 "Characteristic Features of drinking water in Tbilisi area and its close regions").

References

1. Alekin OA (1970) Bases of hydrochemistry. Hydrometeorological Publishing House, Leningrad
2. Drinking water quality parameters. www.water.ru
3. Eristavi V, Danelia A et al (1985) Sources of environment pollution and technical activities of their extermination. Ganatleba, Tbilisi
4. Ghogheliani LD et al (2003) The ecochemical and sanhygienic investigation of the water-reservoir of Zhinvali and Tbilisi. Soros Educ J 5–18
5. Mikadze I (2006) Ecology, Tbilisi
6. Mtsariashvili LA, Khikhadze NV, Kekelidze NP, Jakhutashvili TV, Tulashvili EV (2010) Radioactivity of tap water of Tbilisi city (Georgia) and estimation of radiological risk for population. In: 10th international multidisciplinary scientific geo-conference and EXPO SGEM 2010 – modern management of mine producing, geology and environmental protection, vol II. Albena Resort, Bulgaria, pp 39–46
7. Normative document (2008) Hygienic requirements to quality of drinking water in centralized system of water-supply, Order of the Georgian minister of labor, health and social care (6.08.01) №297/n
8. Supatashvili GD (2003) Hydrochemistry of Georgia (fresh waters). TSU, Tbilisi
9. Technical Requirements to Quality of Drinking Water (2007) Order of the Georgian minister of labor, health and social care (17 Dec 2007) №349/n

Chapter 3
Climate Change Impacts on Water Resources Management with Particular Emphasis on Southern Italy

Michele Vurro, Ivan Portoghese, and Emanuela Bruno

Abstract A methodology to use climate change information in water resources evaluation is developed through a meaningful case study in southern Italy (the Apulia region). The problem of the effective information of climate model simulations with respect to small scale impact studies is developed taking into account the limited predictive capability of climate models. Therefore downscaling and bias-correction requirements are treated through a specific methodology based on a quantile variable correction adopting ground based observation of climate variables. The meteorological forcing for the impact study are obtained through the downscaling of atmospheric variables produced by a Regional Climate Model (RCM) called Protheus. The impact assessment on the water balance of the Apulia region (southern Italy) revealed a marked increase in the variability of hydrologic regimes (both runoff and groundwater recharge) as consequence of the increased rainfall variability predicted for the twenty-first century, while preserving a decreasing in the annual trend. Moreover, the analysis of climate change effects was performed focusing on the rainfall-discharge process of a strategic karst spring supplying the Apulia aqueduct. In this case study, no substantial variations in the annual mean discharge are recognized, although a marked decrease in the mean monthly discharge was found between October and December, which represent the start of the recharge period of Apennine aquifers. Such results represent a crucial water management issue that has to be addressed in terms of adaptation to meet future water resources requirements.

M. Vurro (✉) • I. Portoghese • E. Bruno
Water Research Institute, National Research Council of Italy (CNR-IRSA), UOS Bari Via F. De Blasio 5, Bari 70132, Italy
e-mail: michele.vurro@ba.irsa.cnr.it

F.F. Quercia and D. Vidojevic (eds.), *Clean Soil and Safe Water*, NATO Science for Peace and Security Series C: Environmental Security, DOI 10.1007/978-94-007-2240-8_3,
© Springer Science+Business Media B.V. 2012

3.1 Introduction

Many semi-arid areas, such as the Mediterranean basin, are particularly exposed to the impacts of observed climate change and are projected to suffer for a decrease in water resources with severe impacts upon environmental quality, economic development and social well-being. In order to evaluate these impacts, we shall start from the global climate change effects on the hydrological system. Currently, climate projections for the twenty-first century are based on simulations from Global Climate Models (GCMs) or Regional Climate Models (RCMs) which simulate the climate response to some predicted changes in anthropogenic greenhouse gas emissions [5]. At the present stage, remarkable research efforts have been addressed to assess the predictability of the climate system by improving the climate model physics, resolution, and parameterization for unresolved processes, which result in the development of high-resolution GCMs and RCMs. Nevertheless, the simulated climate behaviour is still far from being consistent at higher frequency and local scales which are basically needed to undertake impact studies trough the implementation of specific hydrological models.

As climate models use crude simplifications of complex atmosphere-land processes, their outputs cannot be expected to exactly reproduce the observed local dynamics. Consequently, basin-scale assessment of climate change impacts may produce large biases in the simulation of river flows if the raw output precipitation from a GCM (or a RCM) is adopted. Particularly, the scale mismatch between climate model output and the spatial resolution (river basin or sub-basin) at which hydrological models are applied (e.g. [4, 18]) is the main limiting factor to the direct use of climate scenarios in impact prediction. Several hydrological impact studies require in fact accurate model simulations not only of time-average conditions but also of the day-to-day (and even sub-daily) variability. In this framework, a rigorous model evaluation of the simulated daily precipitation statistics is an important step in assessing the models' reliability for climate impact applications [9]. Despite the physical consistency of the dynamical downscaling approach, in most cases the bias inherited from the driving GCM inevitably remains (e.g. see [10, 19]). Various additional downscaling techniques are needed to bridge the scale gap between climate model simulation and local scale information needed to force hydrological models of any kind. These techniques range from more simple ones that use trends in climate variables from GCMs simulation to force the historical climate records, to the statistical transfer functions linking local climate to the output of a GCM or RCM (e.g. [17]), to those classified as climatic analogue-procedures.

In this study, an integrate approach to assess the hydrological impact of climate change at regional and local scale is developed including the issues of: (i) climate models to simulate the climatic effects of increasing atmospheric concentration of greenhouse gases; (ii) statistical downscaling techniques to link climate models and hydrological models; (iii) hydrological models at basin scale to simulate hydrological impacts of changing climate; (iv) adaptation measures in order to reduce the impact of climate change on water resource management.

3.2   Background to the Study Region

As typical example of semi-arid Mediterranean area a south-eastern coastal region of Italy (Apulia, Fig. 3.1) was chosen to develop an impact assessment of climate change on water resources. In the past few decades this area has been exposed to a sequence of prolonged droughts, which caused a general decrease in water supply and an increase in demand for irrigation. Moreover, in the past decade the region has been recognized as being among those at the highest risk of desertification in Europe, due to the observed climatic trends and recently intensified agricultural practices.

The climate is markedly Mediterranean, with mild wet winters and hot dry summers (the coldest month is January and the warmest is July). Climate variables, and rainfall in particular, exhibit a marked inter-annual variability which makes water availability a permanent threat to the economic development and ecosystem conservation of the region. In addition, rainfall has also experienced a declining trend, on average, over the past four decades [14]. Due to the main carbonate nature of rocks (high substrate permeability and infiltration of rainwater), the region is poor in rivers and generally in surface water. The most important river basins are located in the

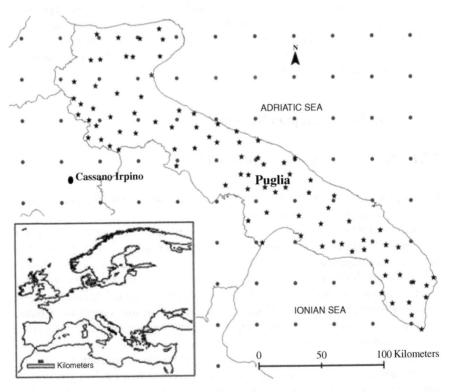

Fig. 3.1 The Apulia case study with precipitation and temperature land measurement network (*black stars*) and RES nodes (*grey circles*)

northern part of the region, where the morphological behavior allows the presence of intermittent rivers. Instead, in the karst area some basins related to a fossil hydrographical network present a superficial flow only during intense events [6, 7].

The region is mainly dominated by agriculture, that is a vital economic resource for the region, with more than 70% of the total area occupied by cropped land. The water resources derived from surface water bodies are limited causing a major constraint to the social and economic development of the region. To overcome this problem, a great aqueduct was built at the beginning of the twentieth century which supplies the region collecting waters coming from some carbonate Apennine springs as Cassano Irpino (mean annual discharge of 2.65 m³/s) and Caposele. Moreover, many conveyance systems were built between 1960 and 1990 to transfer water from the bordering regions in order to supply water for agricultural uses, thus making the region very much relying on the external water resources. This infrastructure is the biggest one for distribution network over Europe. Furthermore, a fast growing trend in the last four decades towards irrigation farming has led to a massive exploitation of groundwater resources. As a result, the groundwater level has dramatically decreased in the river plain aquifers while sea water intrusion is observed in most of the coastal zones [13].

It is therefore quite obvious how crucial is to investigate the possible impacts of climate projections in such hydro-climatic context. Agriculture, water supply and tourism are sectors vulnerable to climate change in the region. In this complex framework, climate change effects on regional water balance and on discharge of Cassano Irpino spring were analyzed.

3.3   Hydrological Impact Models

3.3.1   G-MAT: Water Balance Model

The adopted model, named G-MAT [15], was proved suitable for the evaluation of hydrological water balance in semi-arid conditions.

This model was originally developed for the sustainability assessment of water resources with particular emphasis on groundwater-dependent regions. It considers the major landscape features that determine the soil water balance such as the vegetation activity through the season and the soil moisture storage and flux processes adopting simple parameterizations. G-MAT yields natural groundwater recharge on a monthly basis, through the distributed application of the soil water balance equation (Fig. 3.2), evaluated as the difference between the inflows (rainfall, irrigation) and the outflows (evapotranspiration, surface runoff).

Another peculiar feature of the adopted model is the capability to estimate monthly irrigation amounts, through the soil moisture deficit method [1], at each representative model unit that is characterized by irrigated farming. This is in fact very crucial for both surface and groundwater bodies in Mediterranean regions since agricultural activities account for about 70% of the total water demand. Furthermore,

Fig. 3.2 G-MAT water balance scheme [15]

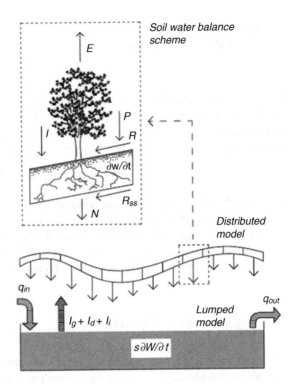

the model parameterization concerning the vegetation cover can be easily modified to investigate the impact of land cover modifications on the hydrological response including water resources allocation for irrigation use.

The spatial resolution of the water balance model is 1 km^2 thus assuring a feasible representation of the spatial heterogeneity of soil, sub-soil, and vegetation features as well as a realistic description of catchment morphology. The monthly time step was chosen as a compromise between the data availability over large domains and the uncertainty introduced by the various data manipulations necessary for the downscaling of climate scenarios through the observation-based corrections.

3.3.2 Spring Discharge Model

To investigate the spring regime, a non-linear rainfall-discharge approach was adopted by using the effective monthly precipitation. Therefore, the hydraulic behavior of a karst basin in the study area was represented by a mathematical model using a non-linear relationship between the estimated recharge and spring discharge records.

Due to the fissured and karst geomorphology [2], watersheds are characterized by a multi-scale spatial heterogeneity. This system can be considered as a non-linear system [12] in which different hydraulic processes take place.

Analytically, the monthly flow rates of Cassano Irpino spring were assessed using the following functions:

$$q_{(t+1)} = k * V_{(t+1)}{}^{\alpha}$$

in which α represents the non-linearity property of the considered system, $q(t+1)$ and $V(t+1)$ represent flow rate and the stored rainfall volume respectively during the month after the considered one. The volumes of effective rainfall stored in the aquifer were calculated by the hydrological water balance equation:

$$V_{(t+1)} = V_t + A*Peff_{(t+1)} - q_t * \Delta t$$

in which Vt and qΔt represents respectively the stored volume and the flow rates of the considered month, Peff(t+1) is the effective rainfall of the month after the considered one and A is the recharge area of the basin. Monthly effective precipitation were computed by pre-processing both observed and RCM rainfall data so as to estimate the evapotranspiration (PET) losses by the Thornthwaite equation [16].

3.4 Meteorological Forcing

The meteorological forcing for the hydrological impact assessments, was obtained though the downscaling of some of the atmospheric variables produced by the RCM model Protheus [3]. The adopted variables at the ground level are air temperature at sub-daily scale, and daily total rainfall. These data were extracted from the climate model runs concerning the control simulation (assuming the ERA-40 1958–1999 dataset as boundary conditions of the regional model), the twentieth century simulation (1953–2000), and the A1b twenty-first century scenario (2001–2050). From the RCM simulation 32 nodes were therefore selected as representative of the entire region (Fig. 3.1). These nodes were associated by proximity rules to the corresponding weather stations scattered throughout the region. Monthly mean maximum and minimum temperatures as well as monthly rainfall totals were considered from the available gauge network (83 temperature stations and 112 rainfall gauge stations).

Due to the raw approximation of land use and topography in the RCM, a statistical downscaling (S-DSC) step was developed to correct the model output for local bias and obtaining realistic meteorological forcing at each of the gauge stations. Since the both impact models adopt a monthly time resolution, the S-DSC method is applied accordingly to the monthly dataset of station records and RCM model simulations, namely mean monthly temperatures and monthly rainfall totals.

Climate model bias can be filtered out following different approaches. In this work the Variable correction method [8] was applied by using a function f(x) based on quantile mapping [19]. Quantiles were estimated for both land reference dataset and RCM control simulations. Their comparison was performed by the quantile-quantile (q-q) plot. The estimated bias was filtered out by using a translation function based on the inverse of the Cumulative Density Function (*cdf*), that allows translating the RCM cdf on the observed one [8, 19]. Let Foj be the cumulative distribution function of the land reference monthly value of a studied variable (in this study, precipitation or temperature) for month j and Fhj be the cumulative distribution function of the corresponding RCM simulation, the corrected ensemble value is obtained through the following equation:

$$Z_j^i = F_{oj}^{-1}(F_{hj}(Y_{ij}))$$

In this way, the *cdf*, and thus the pdf of the post-processed simulation is exactly the same as the *cdf* or pdf of the reference. The q-q correction function obtained from systematic bias of RCM will preserve dynamics and predictions of GCM when applied to some scenario simulation. The local projection of climate scenarios through an intermediary unbiased algorithm yields a robust relative interpretation of climate alteration thus minimizing the reproduction of large model bias into the hydrological model adopted for the impact assessment.

3.5 Water Resource Assessment of the Apulia Region for the Twenty-First Century

3.5.1 Water Balance Projections

The climate forcing obtained from downscaling procedure for each station were interpolated by using the inverse distance weight (IDW) algorithm, in order to obtain monthly input maps for the water balance calculations at a resolution scale of 1 km^2. In particular, a two-steps method based on the correlation between temperature measurements and station elevations was used to interpolate temperature maps, which were then used to estimate potential evapotranspiration through the Hamon model [11].

The climate maps were the used to feed G-MAT model in order to assess monthly runoff, groundwater recharge and irrigation under current and climate change condition. The model parameterizations concerning vegetation, soil and geo-morphology were adopted as invariant throughout the water balance simulation. For the comparative analysis of the water balance signatures between past and scenarios simulations, the historical climate forcing were adopted as reference for the twentieth century. This choice is suitable to take into account the reliability of water resource under present hydrological condition, due to the non-linear solution of soil infiltration and moisture-dependent evapotranspiration and the threshold-based runoff mechanism. The inter-annual variability of the water balance and the driving forcing reconstruction for the historic period and the twenty-first century are shown in Fig. 3.3. Due to the substantial difference between the observation-based simulation of the regional water balance and the twenty-first century simulation, separate annual trends were used for these two subsets to highlight both the historical and

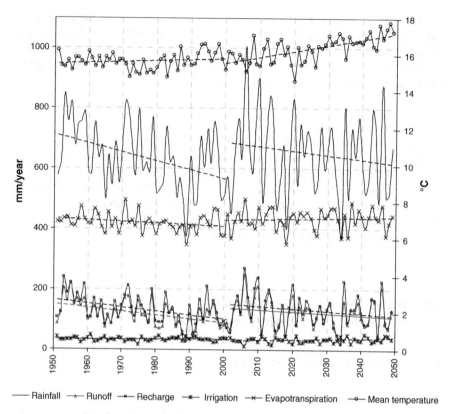

Fig. 3.3 Annual patterns of water balance components for the Apulia region

projected trends of all water balance components. The spatial interpolations of monthly temperature observations and downscaled the A1b twenty-first century scenario reveal an increasing trend of mean temperature due to a rising in both minimum and maximum temperature, while in the historic period only the minimum temperatures were observed to increase. An increasing in the inter-annual variability of precipitation is projected, and thus an increased standard deviation in the annual values is predicted, while a decreasing trend is preserved. Consequently, runoff and groundwater recharge, which are controlled by rainfall regime, are similarly projected into the twenty-first century with an increased variability, while preserving a decreasing trend. The increased variability of precipitation in the twenty-first century is intrinsically related to the dynamic of the global circulation model used as lateral boundary condition to the RCM model which is preserved by the downscaling algorithm. It is therefore arguable that the downscaled rainfall projections suffer from some degree of inconsistency inherited from the GCM and that both river runoff and groundwater recharge may be more rapidly decreasing in the twenty-first century, somehow in agreement with the trend observed in the second half of the twentieth century.

Major water balance variables such as evapotranspiration - counting for more than 60% of the mean annual rainfall - and irrigation show a non-significant trend throughout the investigated periods, thus suggesting a minor sensitivity of the simulated evapotranspiration processes to the projected perturbations in rainfall and temperature as a consequence of the soil moisture capacity which acts as a limiting factor of plant transpirable water.

Related to the Apulia case study, some general remarks can be summarized. Suitable mathematical models are need to represent the non-linear processes such as the unsaturated soil processes, including the role of vegetation coverage, which dominate hydrological predictions in Mediterranean catchments. Then, the downscaling and bias correction issues have to be thoroughly developed in order to obtain suitable information from climate models to be used in the impact analysis on water resources.

According to adopted climate model simulations, the assessment of water resources availability in a water-scarce region has highlighted some peculiar response of streamflow and groundwater recharge to the predicted temperature and rainfall alterations. In particular, no specific trends were detected in the annual water balance components, but a marked increase in the variability of the system as a result of the increased rainfall variability predicted for the twenty-first century. The increased variability of the available water resources is even more severe under the point of view of drought occurrence (and conversely of extremely wet years). Drought events of given return period – and conversely extremely wet event – are therefore expected to be more severe in terms of deviation from the mean values, which remain substantially unchanged between the historic and scenario periods. The increased variability of the hydro-systems in the Mediterranean is therefore confirmed as one of the main water management issue in the near future.

Finally, it has to be concluded that the prediction of hydrological impacts at the catchment scale is still a highly uncertain practice due to the plurality of climate scenarios depending of different models and unpredictable evolution of socio-economic behaviour which strongly influence the water policy in Mediterranean contexts.

3.5.2   Projected Discharge of Cassano Irpino Spring

The hydrological model was calibrated in the period 1981–1986, in which the simulated discharge rates were consistent with the available measurements, and the agreement between observed and modeled time series is reported in Fig. 3.4. The results of model validation for the entire period 1970–1999 are reported in Fig. 3.5. The correlation index between the simulated monthly discharge and the measured discharge of Cassano Irpino spring for the period 1981–1999 is satisfactorily high (0.7). Furthermore the mean monthly prediction error (percent error) of the hydrological model was assessed to range between 0.22% in April and 4.68% in November (Fig. 3.6). To simulate the projection of spring discharge over the period 2000–2029, the hydrological model was forced with the downscaled output from Protheus

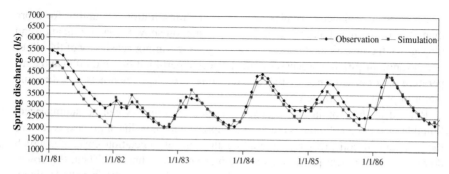

Fig. 3.4 Hydrological model calibration. Spring hydrographs related to observed and simulated monthly discharge related to calibration period (1981–1986)

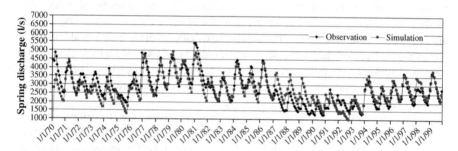

Fig. 3.5 Hydrological model validation. Spring hydrographs related to observed and simulated monthly discharge for the period 1970–1999

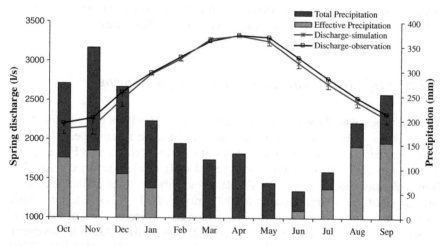

Fig. 3.6 Mean monthly error (*percent error*) of the hydrological model related to 1970–1999 period

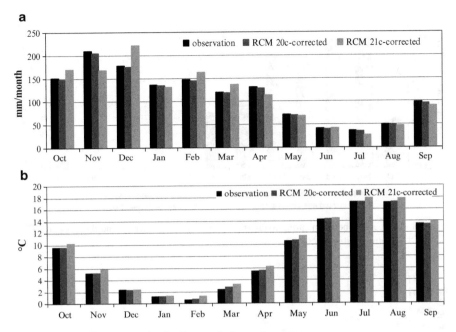

Fig. 3.7 Local climate scenarios for Cassano Irpino spring

System [3]. A control run spanning over 30-year period (1970–1999) and one covering the first part of twenty-first century (2000–2029) under SRES scenario A1b were considered for precipitation and mean air temperature.

The study area refers to one model cell covering about 900 km². As a validation of the downscaling procedure, the comparison between mean monthly values of observations and corrected RCM data over the 1969–1999 are shown in Fig. 3.7a, b. Furthermore, by using the same algorithm, the RCM projections for the twenty-first century were downscaled and the obtained local scenarios were compared with observation and RCM data for twentieth century. In the October-March period, an increase of monthly mean precipitation was projected to range from +13% in February to +27% of December, with an exception for November (−18%). On the contrary, in the remaining part of the hydrological year, a decrease was assessed varying from −2% in May to −24% in July, excepting June (+5%). Meanwhile, concerning mean monthly temperatures, a general increase of 0.5°C in average over the first part of twenty-first century (2000–2029) was obtained from the downscaling procedure (Fig. 3.7b).

To assess the impacts of climate change on the hydrological regime of the Cassano Irpino spring, the downscaled climate variables were used to force the non-linear hydrological model. In order to define the confidence level of the projected scenario for the spring regime, the bias between the observed hydrographs and those ones obtained from corrected RCM data over twentieth century (Fig. 3.8) were assessed as total percent error (due to hydrological model, downscaling procedure

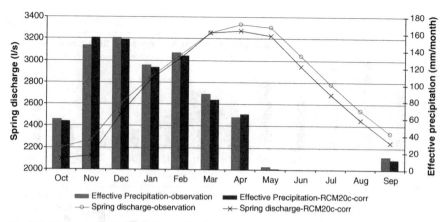

Fig. 3.8 Comparison between spring hydrographs coming from observed and downscaled RCM data over the period 1969–1999

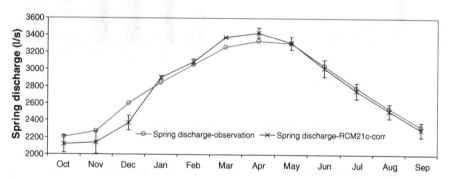

Fig. 3.9 Spring hydrograph comparison between twentieth century (Observed and downscaled RCM run) and twenty-first century (downscaled RCM run)

and climate model). The monthly total error ranged between 0.11% in March and 6% in November.

By analyzing the spring discharge averaged over 30-years (Fig. 3.9), no substantial variations in the annual mean discharge were recognized, although a marked decrease in the mean monthly discharge was found between October and December, which represents the start of the recharge period for Apennine aquifers. The projected local scenario of spring regime (2000–2029) was compared with the observation (1961–1999) in terms of monthly means over 30-years also considering the total error estimation.

The decreased discharge in the late autumn and early winter could be very critical in terms of water resource management, as this type of water supply does not provide any storage system. For this reason, the seasonal discharge reduction is not balanced by any further increase in the following period. To give an idea of the impact of such discharge reduction on the water supply, the spring yield decrease in December (−233 l/s) could be translated into about 100.000 users not being supplied.

3.6   Conclusions

Suitable model interfaces are necessary to couple climate simulations to spatially distributed hydrological models which are used to predict the hydrological response. These model interfaces have to be specifically tailored to the space-time scale and the dominant processes under investigation. At the scale of medium-small size basins, climate change impact investigations require a higher detail in the atmospheric forcing produced by climate models. Despite the progress in climate modelling, in fact, the simulated climate behaviour is still far from being consistent at higher frequency and local scales which are basically needed to capture basic hydrological processes. To overcome the persistent model bias of global and regional climate simulations some kind of post-processing of the atmospheric output has to be applied through an appropriate assimilation of climate observations at finer scale.

A scheme for the assimilation of different data sources into the model prediction chain was represented in this paper enabling to accommodate climate model output with the resolution scale of a generic hydrological impact model. An application of this procedure was presented with regard to a meaningful case study in southern Italy (the Apulia region). Monthly mean temperatures and rainfall totals were considered from the available gauge network to develop a local downscaling starting from Protheus model. Monthly interpolation maps were obtained for temperature and precipitation and used to force a water balance model for the simulation of surface runoff, groundwater recharge and irrigation demand. The increase in mean annual temperature and the enhanced rainfall variability according to the developed local climate scenario has highlighted a marked increase in the variability of water resources availability and irrigation demand in a water-scarce region. Similar results were found on a carbonate aquifer in which a large Apennine spring was selected as a significant hydro-geological system due to the limited anthropogenic disturbance in the recharge areas. A mathematic model to simulate the rainfall-discharge response of this spring was coupled to output of the Protheus model after the aforesaid downscaling procedure. The hydrological projection was in agreement with an increased seasonal variability of the discharge regime with a reduction of water availability in the October-December period.

The increased variability of the hydro-systems in the Mediterranean was therefore confirmed as one of the main water management issue in the near future. Variability has important implications for water management, as it enhances the risk in estimating the true opportunity cost. In other words, the water resource variability requires systematic contingency planning to mitigate drought impacts such as the development of expensive storage capacity to compensate for seasonal and inter-annual flows.

As irrigation is by far the largest water user in the Mediterranean regions, increasing the efficient use of water is a key non-structural approach to water resources management. It is therefore advisable to put more attention on demand management measures rather than building new expensive facilities.

Acknowledgments This research was undertaken under the European Union funded project CIRCE (FP6 Project No. 036961).

References

1. Allen RG, Pereira LS, Raes D, Smith M (1998) Crop evapotranspiration guidelines for computing crop water requirements, Food and Agriculture Organization Irrigation and drainage paper 56, Roma, 300 pp
2. Atckinson TC (1977) Diffuse flow and conduite flow in limestone terrain in Mendip-Hills, Somerset (Great Britain). J Hydrol 35:93–110
3. Artale V, Calmanti S, Carillo A, Dell'Aquila A, Herrmann M, Pisacane G, Ruti PM, Sannino G, Struglia MV, Giorgi F, Bi X, Pal JS, Rauscher S (2009) An atmosphere-ocean regional climate model for the mediterranean area: assessment of a present climate simulation. Clim Dyn. doi:10.1007/s00382-009-0691-8
4. Burlando P, Rosso R (2002) Effects of transient climate change on basin hydrology. 1. Precipitation scenarios for the Arno River, central Italy. Hydrol Process 16:1151–1175
5. Cubasch U, Waszkewitz J, Hegerl G, Perlwitz J (1995) Regional climate changes as simulated in time-slice experiments. Clim Chang 31:273–304
6. De Girolamo AM, Limoni PP, Portoghese I, Vurro M (2001) Utilizzo di Tecniche GIS per la Valutazione del Bilancio Idrogeologico. Applicazione della Metodologia alla Penisola Salentina. L'Acqua 2:57–70
7. De Girolamo AM, Limoni PP, Portoghese I, Vurro M (2002) Il bilancio idrogeologico delle idrostrutture pugliesi: sovrasfruttamento e criteri di gestione. L'Acqua 3:33–45
8. Déqué M (2007) Frequency of precipitation and temperature extremes over France in a anthropogenic scenario: model results and statistical correction according to observed values. Glob Planet Chang 54(1–2):16–26
9. Frei C, Christensen JH, Déqué M, Jacob D, Jones RG, Vidale PL (2003) Daily precipitation statistics in regional climate models: evaluation and intercomparison for the European Alps. J Geophys Res 108(D3):4124–4143
10. Fowler HJ, Blenkinsopa S, Tebaldib C (2007) Review linking climate change modelling to impacts studies: recent advances in downscaling techniques for hydrological modelling. Int J Climatol 27:1547–1578
11. Hamon WR (1963) Computation of direct runoff amounts from storm rainfall. Int Assoc Sci Hydrol Publ 63:52–62
12. Labat D, Manginb A, Ababoua R (2002) Rainfall–runoff relations for karstic springs: multifractal analyses. J Hydrol 256(3–4):176–195
13. Masciopinto C (2005) Pumping-well data for conditioning the realization of the fracture aperture field in groundwater flow models. J Hydrol 309(1–4):210–228
14. Polemio M, Casarano D (2004) Rainfall and drought in Southern Italy (1821–2001). In *The basis of Civilization – Water Science?* IAHS Pub 286:217–227
15. Portoghese I, Uricchio V, Vurro M (2005) A GIS tool for hydrogeological water balance evaluation on a regional scale in semi-arid environments. Comput Geosci 31–1:15–27
16. Thornthwaite CW (1948) An approach toward a rational classification of climate. Geogr Rev (Am Geogr Soc) 38(1):55–94
17. Wilby RL, Wigley TML, Conway D, Jones PD, Hewitson BC, Main J, Wilks DS (1998) Statistical downscaling of general circulation model output: a comparison of methods. Water Resour Res 34:2995–3008
18. Wilby RL, Hay LE, Gutowski WJ Jr, Arritt RW, Takle ES, Pan Z, Leavesley GH, Clark MP (2000) Hydrological responses to dynamically and statistically downscaled climate model output. Geophys Res Lett 27:1199–1202
19. Wood AW, Leung LR, Sridhar V, Lettenmaier DP (2004) Hydrologic implications of dynamical and statistical approaches to downscaling climate model outputs. Clim Chang 62:189–216

Chapter 4
A Geochemical Assessment of Surface Water Quality as a Tool for Indication of Geogenic and Man-Made Constituents of Pollution

Marine A. Nalbandyan

Abstract This article deals with a study of River Hrazdan basin. We studied underlying man-made factors which impact on formation of quality composition of river water. Guided by the proposed approach, we implemented a comparative analysis between – different by character – loads sections of the watershed and background section where man-made factors are either missing or are poorly manifested. On the basis of landscape and geochemical specificity of the study section of the watershed, the article considers natural and man-made associations of heavy metals. The quantity and quality series of the geochemical stream have been studied and collated. Analysis of contents of several heavy metals in underground waters, which are used as a potable water to the city, have been carried out. Assessments of mineral composition in river and underground water have been implemented.

4.1 Introduction

Hydro-geochemical investigations of surface waters produce information which helps to disclose the character and levels of impact of landscapes, to assess pollution levels and to indicate dominating surface water pollutants. Besides, such investigations may be a tool when assessing separate pollution sources. Detailed investigations allow indicating both natural (geogenic) and man-made constituents of pollution. Such an approach may be applied when investigating surface waters in regions exposed to different levels of man-made impacts. Surely, such investigations should be implemented with regard to quality composition of groundwater that feed surface

M.A. Nalbandyan (✉)
The Center for Ecological-Noosphere Studies, The National Academy of Sciences,
Abovyan str., 68, Yerevan, 0025, Armenia
e-mail: marinen3@yahoo.com; ecocentr@sci.am

F.F. Quercia and D. Vidojevic (eds.), *Clean Soil and Safe Water*, NATO Science for Peace
and Security Series C: Environmental Security, DOI 10.1007/978-94-007-2240-8_4,
© Springer Science+Business Media B.V. 2012

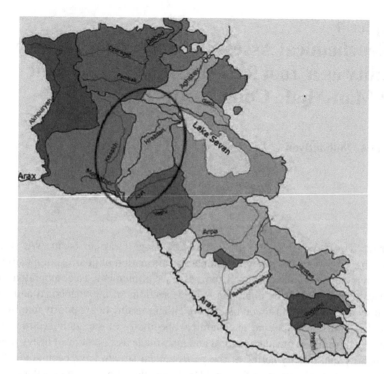

Fig. 4.1 Map of Armenia and location of the studied river basin

streams, emphasizing the share of underground constituent in total feeding. In this regard investigations and quality control of groundwater are of special value.

It is noteworthy that drinking water sources in Armenia are groundwater; therefore studying the system surface waters – groundwater is becoming essential and prospective in terms of controlling and managing fresh water quality in the country.

We employ a method of geochemical assessment of surface waters emphasizing river waters in the study of water composition of natural and man-altered landscapes. For indicating geogenic and man-made constituents of pollution one should first investigate into the geological structure of the river watershed, groundwater quality and level, a range of geographical factors (site steepness, density of forests, soils etc.), levels of man-made impacts (urbanization level, character and scale of operating plants, level of agricultural development) and so on.

This article deals with a study of River Hrazdan basin (Fig. 4.1). The river basin hosts about 70% of Armenia's industrial enterprises and municipal services of Armenia's capital – the city of Yerevan – and of the cities of Sevan, Hrazdan, Charentsavan and Abovian, as well as an impressive number of smaller settlements.

On the basis of landscape and geochemical specificity of the study section of the watershed, the article considers the formed natural and man-made associations of heavy metals. We studied underlying man-made factors which impact on formation of the quality composition of river water. Guided by the proposed approach, we implemented a comparative analysis between different by character and exposed to different loads sections of the watershed and its background section where man-made factors are either missing or are poorly manifested.

4.2   Materials and Methods

The article highlights the outcomes of monitoring surveys implemented between years 2004 and 2006.

Samples were collected, conserved, transported and stored following Standard Operational Procedures (SOPs) which were developed following the International Standardization Organization (ISO) methods (ISO-5667-1, ISO-5667-2, ISO-5667-3, 1998) [6].

Field measurements were carried out on a monthly basis. In-situ measurements included hydrogen index (pH), conductivity, turbidity, dissolved oxygen, temperature and salinity and were done applying a portable analyzer U-10 (Horiba). Water discharge was measured on a USGS-type AA Current Meter using a Data Storage Computer of AquaCalc 500 model (Rickly Hydrological Co).

Common ions Na, K, Mg, Ca, SO_4, Cl, HCO_3, CO_3 were studied on a quarterly basis and were determined in non-acidified and filtered samples. The quantity of ions was determined in lab conditions on a SF-46, a portable spectrophotometer DR/2400 Hach. On a monthly basis soluble forms of the following heavy metals (HMs) were determined: Cu, Mo, Zn, Cr, Ni, Mn, Cd, As, Hg, Co, Pb, Ag. Samples for determination of mercury were collected into a glass container, acidified by HNO_3 until pH < 2 was reached, and then $K_2Cr_2O_7$ was added. While determining concentrations of other HMs, the samples were filtered through membrane filters with a pore diameter 0.45 μm, then acidified by HNO_3 (1:1) until reaching pH ~ 2 and stored in polyethylene containers. HMs were determined on a PE Analist 800 through the atomic-absorption method with graphite atomizer, and flame photometry. Analyte concentrations were measured following the developed ISO-based SOPs [3, 4].

To characterize the level of anomaly of chemical element distribution in studied media, we applied an anthropogenic Concentration Coefficient (Cc). It represents the relation of element contents in an anomalous study object (Ci) with respect to its background contents (Cb): Cc=Ci/Cb. To obtain the integral characteristic of poly-element pollution, a Summary Concentration Index (Zc) was calculated, which represents the sum of background-standardized contents of elements in the sample [7]: Zc=Σ Cc − (n−1), where n is the number of elements with Cc > 1.

Data on monthly river water quality monitoring at stations Marmarik –Aghavnadzor (st.0), Hrazdan-Masis (st.8) and at 7 stations within the border of the city of Yerevan were used (Figs. 4.2 and 4.3).

Data on River Hrazdan were obtained as result of the monitoring implemented in the frame of a NATO SfP project "South Caucasus River Monitoring" (2002–2008). This research was reported also in articles from [8, 9].

Fig. 4.2 A schematic
map of studied areas
(BP-1 – natural area,
BP-2 – basin area within
the boundaries of the city
of Yerevan, BP-3 – basin's
area where agricultural
activities dominate)

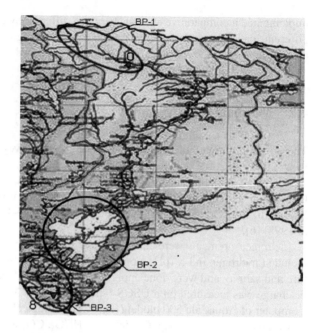

4.3 Physico-geographical, Economic and Hydrological Characteristics of the River Hrazdan Basin

In the Republic of Armenia River Hrazdan represents a strategic water artery which needs a serious control of the quality and management of its ecosystem. The watershed of River Hrazdan is a complex ecosystem, which experiences heavy loads of different man-induced factors. The latter include sources of industrial and agricultural pollution, as well as from domestic sewage.

River Hrazdan is the longest tributary of River Araks in the territory of Armenia. The length of River Hrazdan is 146 km and its watershed area is 2,560 km^2 (excluding Lake Sevan basin). The mean long-term water discharge is 7.57 m^3/s. The Hrazdan river originates from Lake Sevan and flows into River Araks. The total drop of the river is 1,090 m and the mean annual runoff is 0.71 km^3.

4.3.1 Characteristics of the Natural Area of the Basin

In the studied area, surface and groundwater source locations are the slopes of Mt. Tsaghkuniats and the Pambaki Range. The basin is composed predominantly by impermeable layers of mountain rocks. The landscapes are typical woody. The most widespread tree species include oak, beech, Caucasian hornbeam.

In that area, 75% of river water originates from rainfall and melted snow while the rest 25% from groundwater. No facilities intended for industrial purposes are found there.

Fig. 4.3 Map of monitoring stations location on River Hrazdan (within Yerevan city borders)

4.3.2 Characteristics of an Area of the Basin Within the Borders of the City of Yerevan

In the studied area, surface and groundwater source sites are mainly the western slopes of the Geghama Range. They display a unique development of the hydrographic network. In groundwater formation and distribution within the boundaries of the city of Yerevan, a notable role is played by rock composition in that region.

A considerable part of western slopes display fractured Quaternary lavas – andesites and andesite-basalts. In those rocks alteration of several layers is observed, which are separated by slag formations. In most cases such layers are water resistant. The lava rock thickness varies between 700 and 800 m [8]. In the whole, rocks are water pervious and water bearing. Sedimentary formations show a limited distribution and are represented there by clay and argillaceous slates and sandstone [2].

In this area 55% of river water is fed from rainfalls and melted snow and 45% from groundwater.

The city's territory is home to a number of industrial enterprises including "Molybdenum Production", "Chistoe zhelezo" plants, "Nairit" (producing artificial resins) and "Polyplast", a thermal energy generation plant.

4.3.3 Characteristics of the Basin's Area Where Agricultural Activities Dominate

In the lower stream the river crosses the Ararat Valley. The geological structure of the Ararat Valley involves lymno-fluvial and effusive water bearing formations which thickness reaches 500 m. Beneath those formations, a folded water resistant formation is found which is represented by Paleozoic and Mezocainosoic sandstone, clayey and carbonate rocks [1].

Eighty percent of springs there are originated from the lavas. The water balance of the Ararat bowl involves groundwaters which differ by conditions of formation, chemical composition and bedding conditions. The basic feeding source to aquifers is atmospheric precipitation and infiltration of surface waters [2].

Water accumulation and flow mostly occurs through andesite-basalts and loose fragmented materials of under-bed Quaternary sediments. Data from numerous wells and geophysical surveys prove that the basin feeding ways mostly coincide with the modern river network [9].

The landscape of the flat portion of the Ararat Valley is desert-semi-desert, with typical grey-brownish soils.

In some areas sandy hills, solonchaks, saline soils as well as wetland coexist. A major part of the plain is cultivated and covered with crops – orchards and wine yards- and irrigation requiring soils. The uncultivated part of the territory is covered with xerophyte, halophyte and wormwood.

In that part of the basin feeding from groundwater amounts to 62.5%, while rain and melted snow contribute for 37.5% to total river water supply.

4.4   Results

4.4.1   Heavy Metals Study

4.4.1.1   Characteristics of Quality Series of Geochemical Streams in Different Sections of the River Basin

We studied and collated quantity and quality series of geochemical streams of HMs in different sections of River Hrazdan watershed. To that purpose coefficients of Cc concentrations and of summary index Zc were calculated.

Coefficients of heavy metal concentrations for the background section were assumed equal to 1 (Table 4.1).

Table 4.1 Heavy metals concentration coefficients for the background section of the watershed

Elements	Cu	Pb	Cr	As	Zn	Mo	Ni	Mn	Cd
Cc	1	1	1	1	1	1	1	1	1

Table 4.2 Heavy metals concentration coefficients for the Yerevan agglomeration section

Elements	Cr	Cu	Pb	Ni	Cd	Zn	Ag	Mn	Mo	Zc
Cc	23.6	17.3	2	9	4	3.3	3	3	2	59.2

Table 4.3 Heavy metals concentration coefficients for the watershed section experiencing man-made farming load

Elements	As	Ni	Hg	Mn	Zn	Cd	Cu	Cr	Mo	Zc
Cc	10.3	5	4	3.6	3.4	2.4	2.3	0.7	0.5	24.2

In Formula (4.1) HMs are arranged in regressive order of their concentrations and value in the natural geochemical setting of the site.

$$Zn_{(1)} - Mn_{(1)} - Cu_{(1)} - Cd_{(1)} - Cr_{(1)} - Ni_{(1)} - Mo_{(1)} - As_{(1)} - Pb_{(1)} \quad (4.1)$$

Coefficients calculated relatively to their background concentrations (Table 4.2) characterize the impact level of man-made load on river water quality within the boundaries of Yerevan agglomeration. This is demonstrated not only by an increase of HMs concentration of geogenic genesis, but also by the evident man-made origin of heavy metals, alien to the natural setting and to the geochemical peculiarities of the site.

In the determined geochemical association (Formula 4.2) dominating metals with a manifold increase in concentrations are Cr, Cu, Ni. The sources of the detectable man-made pollution are industrial enterprises, which waste waters are not subject to any treatment before emptying into the river.

$$Cr_{(23,6)} - Cu_{(17,3)} - Ni_{(9)} - Cd_{(4)} - Zn_{(3,3)} - Ag_{(3)} - Mn_{(3)} - Pb_{(2)} - Mo_{(2)} \quad (4.2)$$

Thus in the given section of the watershed a qualitatively different man-made association of HMs has been determined which allows the assessment of the value of man-made constituents in the general geochemical series.

In the lower reaches of the river, the watershed is covered with farmlands in which vegetables and cereals are grown. Vast areas there are allotted to vineyards. In the studied river section, a quantity and quality change in the geochemical association (Formula 4.3) is observed. Table 4.3 gives coefficients of concentrations of HMs typical of this section of the watershed.

$$As_{(10,3)} - Ni_{(5)} - Hg_{(4)} - Mn_{(3,6)} - Zn_{(3,4)} - Cd_{(2,4)} - Cu_{(2,3)} - Cr_{(0,7)} - Mo_{(0,5)} \quad (4.3)$$

The analysis of geochemical series allowed the determination of new metals (As, Hg) which are not typical of other sections of the watershed. Their rather high

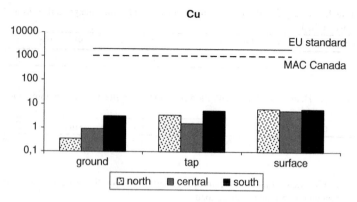

Fig. 4.4 Dynamics of *copper* contents (mkg/L) in different parts of Yerevan for groundwater, tap and surface waters

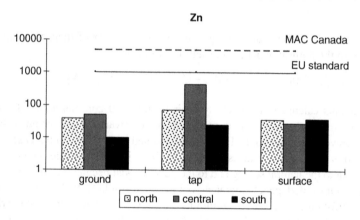

Fig. 4.5 Dynamics of zinc contents (mkg/L) in different parts of Yerevan for groundwater, tap and surface waters

concentrations, in water result from application of pesticides to cultivated land. However, lower concentrations of Ni, Cd, Cu have been observed as compared with the previous section of the watershed, this evidencing the absence of industrial pollution sources.

4.4.1.2 Assessing the Quality of Surface and Potable Waters Within the Boundary of Yerevan Agglomeration

This chapter gives a comparative analysis of contents of several HMs in groundwater which is used as water supply source to the city and contents in River Hrazdan waters. A comparison was carried out between the quality of groundwater from wells and that of the same water distributed as tap water to the consumer.

The results of the analysis indicate that the contents of the considered HMs in tap water are higher against that in water directly pumped from the well (Figs. 4.4–4.6).

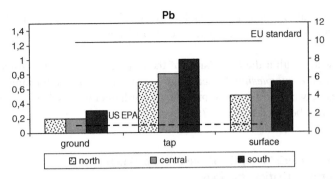

Fig. 4.6 Dynamics of lead contents (mkg/L) in different parts of Yerevan for groundwater, tap and surface waters

Concentrations of Pb and Cu were notably higher in the river waters as compared with groundwaters, and those of Zn were almost equal. Concentrations of Pb and Zn in tap water were higher as compared to groundwater and river water. According to data, tap water contained higher concentrations of Zn in the center of the city vs. the suburbs, whereas Pb was relatively high in the southern part of the city.

A further assessment was performed by comparing HM contents to EU, MAC (Maximum Acceptable Concentration), Canada and US EPA standards. Consistent with assessment by European and Canadian standards, concentrations of the studied metals do not exceed allowable limits. Pb exceeds the limit given by US EPA standards.

4.4.2   Study of Mineral Compositions

4.4.2.1   Characteristic of the Background Section of the Basin

The water from the given river section is characterized as hydro-carbonate-sulfate, with low mineralization. According to estimates, pH varies between 7.8 and 8.0. Mean total annual mineralization is 208.24 mg/L. Carbonate hardness is equal to total hardness and its average value is 1.5 mg-eqv/L. According to classification of natural waters, the background section basin water is characterized as very soft [5].

4.4.2.2   Characteristics of the Basin Section Within the Boundaries of the City of Yerevan

The water from this section of River Hrazdan is characterized as *hydro-carbonate-sulfate with relatively high mineralization*. According to estimates, pH varies between 6.6 and 7.8. Mean total annual mineralization is 556.1 mg/L. Carbonate hardness is equal to total hardness and its mean value is 3.2 mg-eqv/L. According to classification of natural waters this section of basin water is characterized *as soft* [3, 4].

Groundwater within the boundaries of the city belong to *the hydro-carbonate-chloride, calcium-magnesium* (or calcium-sodium) type. The total mineralization value varies between 170 and 470 mg/L, total hardness range is 1.0–4.0 mg-eqv/L, while pH varies between 6.25 and 7.3 [8].

4.4.2.3   Characteristics of the Basin Section Where Farm Activities Dominate

The river water in this section of the basin is characterized as *hydro-carbonate-chloride*. According to estimates, pH varies between 7.1 and 7.3. Meanly, total annual mineralization makes 749.7 mg/L. Carbonate hardness is equal to total hardness and averages 6.3 mg-eqv/L and consistent to classification of natural waters is characterized as water of *medium hardness* [5].

Groundwater there belongs *to hydro-carbonate-chloride-sulfate (mostly calcium-sodium-magnesium)* category. Total mineralization varies between 406 and 566 mg/L, and pH between 6.3 and 6.58 [9].

Collation the separately considered sections of the basin one may then conclude that an increase in water mineralization and water hardness level is detected from the river head to the mouth.

It should be noted that increase in mineralization level in the urban territory is determined by man-made impacts of waste waters, whereas in the lower reaches section such an increase is mainly induced by domination of highly mineralized groundwater feeding the river.

4.5   Conclusion

Investigations of hydro-geochemical peculiarities of River Hrazdan basin supported the indication of geogenic and man-made constituents of pollution. The assessment of River Hrazdan water quality in different sections of the basin allowed the detection of dominating HMs in the series of man-made geochemical associations.

Salt composition peculiarities together with groundwater and surface water mineralization levels may also be a tool for assessing man-made impacts on water quality. Comparison between the quality of River Hrazdan and groundwater within the boundaries of its watershed, this research defined sections experiencing maximal anthropogenic impacts.

Decay of the quality of tap water within the city of Yerevan is not a source-related concern, but rather a technical problem; the level of water quality safety when supplied from water source to the consumer may be estimated as unsatisfactory.

Acknowledgments This research was based on data from international NATO/OSCE project N 977991 "South Caucasus River Monitoring", 2002–2008. The author wishes to express special thanks to Lilit Grigoryan for technical assistance.

References

1. Amroyan AE, Harutyunyan RG, Orbelyan ES (1974) The Ararat bowl. In: Geology of the ArmSSR, vol 8, Gidrogeoekologia. AS ArmSSR, Yerevan, pp 227–245
2. Avetissyan VA, Balyan SP, and others (1974) The Geghama Range. In: Geology of Armenian SSR, vol VIII, Hydro-ecology. Erevanski Gosudarstweni Universitet, Yerevan, pp 130–150
3. Fomin GS (2000) Water: control of chemical, bacterial and radiation safety by international standards. Standard operational procedures for determination of basic ions and biogenic elements in surface waters, ISO-6878, ISO-9297, ISO-7150-1, ISO-7890-3, ISO-6777, ISO-9963-1, ISO-9964-3, 6058, 6059, 9280, Encyclopedical handbook. Protector, Moscow, p 836 (in Russian)
4. Fomin GS (2000) Water: control of chemical, bacterial and radiation safety by international standards. Standard operational procedures for determination of heavy metals in surface waters, ISO-8288, ISO-9174, ISO-5961, Encyclopedical handbook. Protector, Moscow, p 836 (in Russian)
5. Koryukhina TA, Churbanova IN (1983) Chemistry of water and microbiology. Stroyizdat, Moscow, pp 78–79
6. Manual for sample conservation and storage. Manual for sampling methods (1998) Standard operational procedures for river water sample collection, conservation and storage, ISO-5667-1, ISO-5667-2, ISO −5667-3. web-site document
7. Perelman AI, Kasimov NS (1999) Landscape geochemistry. Visshaya Shkola, Moscow, p 768
8. Shaginyan GV (2005) On hydro-chemical characteristics of some potable water supplied to Yerevan. Izv NAS Armen Earth Sci VIII(1):41–45
9. Shaginyan GV, Khalatyan ES, Kjuregyan TN (2004) Hydrochemistry of potable waters in southern districts of Yerevan. Izvdf NAS Armen Earth Sci VII(1):49–54

Chapter 5
Risk-Based Approach to Contaminated Land and Groundwater Assessment: Two Case Studies

Eleonora Wcisło, Marek Korcz, Jacek Długosz, and Alecos Demetriades

Abstract The human health risk-based approach to contaminated land and groundwater assessment is described. It was developed under the EC financed NORISC project, and later on included in the technical guidelines for guiding remedial activities in Cyprus and Poland. The approach is designed as a two-step process: (1) preliminary site assessment and (2) site-specific assessment. The proposed site-specific human health risk assessment (HRA) process consists of two key phases: (a) baseline human health risk assessment (BHRA), including development of a data set, exposure assessment, toxicity assessment and risk characterisation, and (b) development of site-specific risk-based remedial levels (RBRLs). The proposed site-specific HRA method is based on the methodology of the United States Environmental Protection Agency (USEPA), and is applied throughout the entire remediation process, including the phases before, during and after remediation. In order to present the practical application of the HRA procedure two case studies are described: (1) Moni industrial site (Cyprus) – concentrating on contaminated soil, and (2) Milan-Meton site (Italy) and Linz-Heilham site (Austria) – assessing groundwater contamination.

E. Wcisło (✉) • M. Korcz • J. Długosz
Institute for Ecology of Industrial Areas, 6 Kossutha Street, Katowice 40-844, Poland
e-mail: wci@ietu.katowice.pl

A. Demetriades
Institute of Geology and Mineral Exploration, Hellas, Athens, Greece
e-mail: ademetriades@igme.gr

F.F. Quercia and D. Vidojevic (eds.), *Clean Soil and Safe Water*, NATO Science for Peace and Security Series C: Environmental Security, DOI 10.1007/978-94-007-2240-8_5, © Springer Science+Business Media B.V. 2012

5.1   Risk-Based Approach Designed for the Assessment of Contaminated Land in Cyprus and Poland

Risk-based assessment of contaminated land is applied throughout Europe and the United States [1–4, 8, 21]. The risk assessment methods, developed by the (USEPA), have been commonly applied since the 1990s to support remedial decisions at Superfund sites [8, 9]. European countries use different risk assessment models, including USEPA ones, as for example the Czech Republic and Poland [6, 7, 22, 23, 26].

The human health risk-based approach to contaminated soil and groundwater assessment was developed under the EC financed NORISC[1] project, and later on included in the technical guidelines for guiding remedial activities in Cyprus and Poland [24, 25, 28]. The approach was designed as a two-step process: (1) preliminary site assessment and (2) site-specific assessment (Fig. 5.1).

For the purposes of preliminary site assessment, maximum contaminant concentrations, detected in soil or groundwater, are compared with the relevant risk-based screening values. When the screening values are exceeded, follow-up site investigations and site-specific human health risk assessment (HRA) are required.

5.1.1   Site-Specific HRA Method

The described site-specific HRA method is based on the methodology of the USEPA, and is applied throughout the entire remediation process, including the phases before, during and after remediation [8–11, 13, 14, 16–19]. The site-specific human health risk assessment (HRA) process consists of two key phases: (a) baseline human health risk assessment (BHRA), and (b) development of site-specific risk-based remedial levels (RBRLs).

To estimate BHRA, the following steps are undertaken:

- development of data set,
- exposure assessment,
- toxicity assessment, and
- risk characterisation.

Soil exposure is assessed under defined present or future land use patterns, e.g., industrial, residential and recreational.

Soil exposure pathways, associated with these land use patterns, include:

- incidental soil and dust ingestion,

---

[1] NORISC (Network Oriented Risk assessment by In-situ Screening of Contaminated sites) – EC FP5 project (2001–2003), (http://www.norisc.com/).

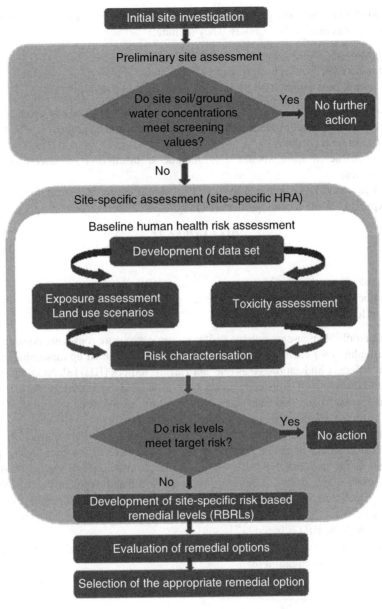

Fig. 5.1 Risk-based approach to contaminated land in Cyprus and Poland (Adapted from [24, 25, 27])

- dermal contact,
- inhalation of fugitive dust, and
- inhalation of volatiles outdoors.

Groundwater exposure pathways are evaluated under the residential scenario, if groundwater is used in households. They include:

- ingestion of groundwater used as drinking water,
- dermal contact while showering or bathing, and
- inhalation of volatiles from groundwater during household use.

Human health risk is characterised separately for carcinogenic and non-carcinogenic effects. Chemicals that produce both non-carcinogenic and carcinogenic effects are evaluated in both groups.

Risk characterisation should also include a discussion on accompanying uncertainties.

Non-cancer risk is calculated for each exposure route (oral, dermal, inhalation), and expressed as the hazard quotient (HQ) [8]:

$$HQ = CDI \, / \, RfD$$

where:

HQ – hazard quotient (unitless),
CDI – chemical daily intake (mg/kg/day),
RfD – reference dose (mg/kg/day).

When multiple non-carcinogens and/or multiple exposure routes are considered, HQs calculated for each chemical/exposure route are summed up (assuming additivity of effects), and expressed as relevant Hazard Indices (HIs) [8]. Next, by summing up HIs across all exposure pathways, the total HI is calculated for the selected land use scenario and relevant receptors (e.g., child, adult). It is not expected that there is any risk of toxic effects when the total HI is less than 1.

If the HI exceeds unity, the compounds are segregated by effects, target organs, and by mechanism of action, and separate HIs are derived for each group [8]. Because of the potential for different health effects/target organs via oral/dermal and inhalation exposures, these exposures are evaluated separately [18]. Non-carcinogenic contaminants are grouped according to primary target organs, i.e., the organs most affected (experience critical effects) by chronic exposure to a chemical, and on which the RfD is based [17].

Cancer risks are quantified for each exposure route (oral, dermal and inhalation) using the following equation [8]:

$$CR = CDI \times CSF$$

where:

CR – cancer risk,
CDI – chemical daily intake averaged over 70 years (mg/kg/day),
CSF – cancer slope factor (mg/kg/day)$^{-1}$.

When multiple carcinogens and/or multiple exposure routes are considered, cancer risks for each carcinogen are added (assuming additivity of effects), and the cancer risk for each exposure pathway is calculated. Summing up CRs across all exposure pathways the total CR is estimated for the selected land use scenario and relevant receptors [17].

Under the scenarios referring to both receptors – a child and an adult (*i.e.*, residential and recreational), cancer risks are calculated separately for these receptors, and then summed up to give the total cancer risk for the aggregate resident/recreational user.

The CRs are compared with an acceptable cancer risk value. It was proposed to establish the value of 1E-06 as the acceptable cancer risk for individual contaminants in site-specific risk assessments, performed both in Cyprus and Poland [24, 27].

Setting up a 1E-06 risk level for individual chemicals and pathways should generally lead to cumulative site risks within the range of 1E-06 to 1E-04 for the combinations of chemicals [13].

5.1.2 Development of Site-Specific Risk-Based Remedial Levels

If risk estimates exceed target risk (TR) (*e.g.*, cancer target risk – CTR = 1E-06 for carcinogens and/or non-cancer target risk – NTR (HQ) = 1 for non-carcinogens), then site-specific risk-based remedial levels (RBRLs) are developed for the purposes of guiding remedial activities.

For supporting the process of remediation of contaminated sites in both Cyprus and Poland a simplified method was proposed for calculating RBRLs according to *USEPA* [16]. This method uses site-specific data under the relevant exposure scenario:

$$RBRL = C \times Target\ Risk\ /\ Calculated\ Risk$$

where:

C – chemical concentration in soil or groundwater
$RBRL$ – risk-based concentration (oral/dermal or inhalation).

RBRLs are calculated separately for carcinogenic and non-carcinogenic effects, and separately for oral/dermal and inhalation exposures, because of the potential for different health effects (target organs) through these routes, as recommended by the *USEPA* [18].

Concerning non-carcinogens, if more than one chemical affects the same target organ or organ system, RBRLs should be adjusted to reflect the potential for additive risks [18]. In such cases, RBRLs calculated for individual non-carcinogens should be divided by the number of chemicals with the same target organs/effects:

$$ARBRL = RBRL\ /\ n$$

where:

$ARBC$ – adjusted risk-based remedial level (for exposure to multiple non-carcinogens with the same target organs/effects),

RBRL – site-specific risk-based remedial level for an individual non-carcinogen
(oral/dermal or inhalation),

n – number of non-carcinogens with the same target organs/effects.

All types of RBRLs, developed for a given contaminant, are compared and the lowest of these values is suggested as a preliminary remedial level ($RBRL_{lowest}$). If $RBRL_{lowest}$ is below the relevant background concentration (BC) for a given contaminant, the RBRL should be established as equal to BC. This is a very important consideration, since the RBRL of any contaminant must not be lower than its local natural geochemical background.

The developed site-specific RBRLs are used next for evaluating remedial options, conducted during the feasibility study. The RBRLs levels combined with the results of the feasibility study help to select the appropriate remedial action for the site (see Fig. 5.1). Additional risk evaluations may also be necessary to be conducted during a remediation phase, and after the remediation is completed to identify residual risks.

5.2 Application of Risk-Based Approach to Contaminated Sites

Application of the risk-based approach to contaminated sites is described by using two case studies:

- Case study 1 – related to contaminated soil at the Moni industrial site, Cyprus (Cyprus project *"National Inventory of Potential Sources of Soil Contamination in Cyprus"* – 2005–2007, Tender 5/2004, Cyprus Geological Survey Department), and
- Case study 2 – related to contaminated groundwater at the Milan-Meton site, Italy and Linz-Heilham site, Austria (INCORE project).[2]

5.2.1 Case Study 1 – Moni Site, Cyprus

The Moni industrial area, Cyprus (Fig. 5.2) was selected as a case study:

- to illustrate the site-specific HRA process and to develop preliminary remedial levels for soil contaminants, and
- to study in practice the variation in the site-specific HRA outputs, depending on the different land uses (industrial, residential and recreational) [29].

The investigation area covered about 160,000 m² of the Moni industrial area.

According to the Cypriot national inventory of potential sources of soil contamination the following industrial activities were identified at the Moni site: non-metallic industries, metal works, tannery and animal rearing.

---

[2] INCORE (Integrated Concept for Groundwater Remediation) – EC FP5 (2000–2003), (http://www.umweltwirtschaft-uw.de/incore/).

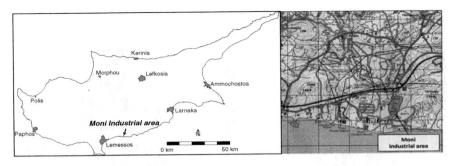

Fig. 5.2 Location of the Moni industrial area, Cyprus

The considered contaminants included: Al, As, Ba, Be, Cd, Co, Cr (VI), Cu, Fe, Li, Mn, Ni, Pb, Sr, Se, Sb and Zn (17 elements).

Exposure was assessed under three potential land use scenarios: residential, industrial and recreational, and by considering three surface soil exposure pathways: incidental soil and dust ingestion, dermal contact with soil, and inhalation of fugitive dust.

For the assessment of site-specific HRA and developing RBRLs for soil contaminants[3] at the Moni site, the upgraded NORISC-HRA software, developed by IETU's team under EU FP5 Project-NORISC, was used [25] by taking into consideration that this application employs the same risk assessment models, as proposed to contaminated land assessment in Cyprus. The NORISC-HRA software package estimates the level and spatial distribution of human health risks at a given site, as well as it sets up RBRLs. Risk results are visualised to assist the decision-making process and communication between different stakeholder groups.

Non-cancer risk and cancer risk values were estimated for each soil sampling point under the considered scenarios. Risks were characterised by two independent sets of risk zones:

- non-cancer risk zones with total HIs exceeding NTR (*i.e.*, HI = 1), and
- cancer risk zones with total CRs exceeding CTR = 1E-06.

The risk zones with risk above and below TR levels were displayed in light and dark grey, respectively. For each established zone, the arithmetic mean values of HQs, HIs or cancer risks were calculated, and a summary of risk assessment results was given.

Figure 5.3 illustrates non-cancer risk zones with total HI > 1 established under the industrial, residential and recreational scenarios. Two non-cancer risk zones with total HI higher than 1 were established under the industrial scenario (adult receptor). Two and one non-cancer risk zones were established under the recreational and residential scenarios for the child receptor, respectively.

---

[3] With the exception of lead; risk associated with industrial and residential exposures to lead in soil was assessed only by comparing lead concentrations in soil with the soil-Pb screening values recommended by USEPA, *i.e.*, 800 and 400 mg/kg for industrial and residential scenarios, respectively [12, 15, 20].

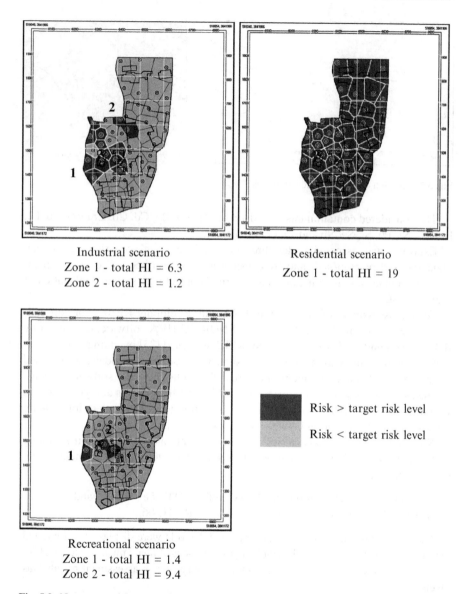

Industrial scenario
Zone 1 - total HI = 6.3
Zone 2 - total HI = 1.2

Residential scenario
Zone 1 - total HI = 19

Risk > target risk level

Risk < target risk level

Recreational scenario
Zone 1 - total HI = 1.4
Zone 2 - total HI = 9.4

Fig. 5.3 Non-cancer risk zones with total HI > 1 under the industrial (adult receptor), residential (child receptor) and recreational (child receptor) scenarios for the Moni site, Cyprus

Under all considered exposure scenarios the total CR exceeded the CTR value of 1E-06 at all sampling locations, establishing only one cancer risk zone (Fig. 5.4). The average total CR, estimated under the residential scenario (1.95E-04) is the highest among all considered scenarios, *i.e.*, almost one order of magnitude higher than the total CR risk, estimated under the industrial scenario (5.10E-05), and two order of magnitude higher than under the recreational scenario (6.70E-06).

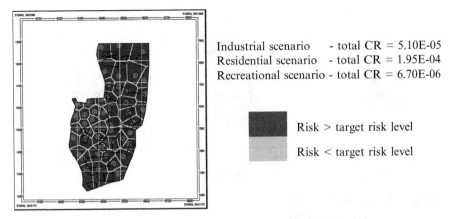

Industrial scenario   - total CR = 5.10E-05
Residential scenario   - total CR = 1.95E-04
Recreational scenario - total CR = 6.70E-06

Risk > target risk level

Risk < target risk level

Fig. 5.4 Cancer risk zone with total CR > 1E-06 under the industrial, residential and recreational scenarios for the Moni site, Cyprus

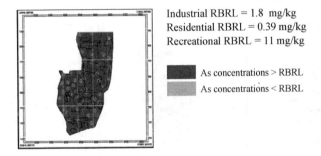

Industrial RBRL = 1.8 mg/kg
Residential RBRL = 0.39 mg/kg
Recreational RBRL = 11 mg/kg

As concentrations > RBRL

As concentrations < RBRL

Fig. 5.5 Arsenic remedial zones under the industrial, residential and recreational scenarios for the Moni site, Cyprus

RBRLs were calculated for all considered soil contaminants detected at the Moni industrial site. If for a given substance the TR was exceeded at least at one point, the relevant point map was displayed. Polygons with estimated concentrations below and above the calculated RBRLs were marked in light and dark grey, respectively. The latter polygons indicated sub-areas for which remediation or risk management procedures should be undertaken (remedial zones). The remedial zones for arsenic, chromium (VI) and antimony are presented on Figs. 5.5–5.7, respectively. They depend on the exposure scenario. Figure 5.8 illustrates remedial zones for aluminium, iron, manganese and selenium under the residential scenario.

Findings of the HRA indicated that:

- Contaminated soil in its present condition might pose potential cancer and non-cancer risks to industrial workers, potential residents and recreational users.
- Under all considered scenarios the total cancer risk exceeds CTR of 1E-06 at all locations.
- The average total cancer risk, estimated under the residential scenario (1.95E-04), is the highest among all considered scenarios.

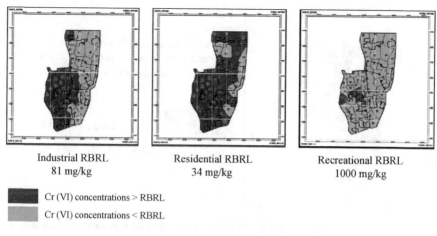

Industrial RBRL
81 mg/kg

Residential RBRL
34 mg/kg

Recreational RBRL
1000 mg/kg

■ Cr (VI) concentrations > RBRL

■ Cr (VI) concentrations < RBRL

Fig. 5.6 Chromium (VI) remedial zones under the industrial, residential and recreational scenarios for the Moni site, Cyprus

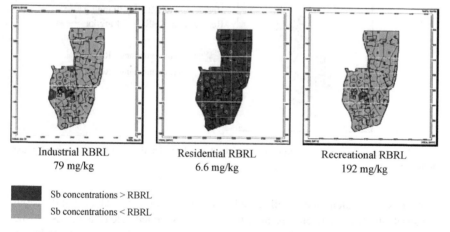

Industrial RBRL
79 mg/kg

Residential RBRL
6.6 mg/kg

Recreational RBRL
192 mg/kg

■ Sb concentrations > RBRL

■ Sb concentrations < RBRL

Fig. 5.7 Antimony remedial zones under the industrial, residential and recreational scenarios for the Moni site, Cyprus

- Under all considered scenarios the main portion of total CR was contributed by Cr (VI) inhalation and As oral risks.
- In order to reduce the health risk to a safe level for potential site users, surface soil should be remediated to the relevant RBRLs, depending on the end land use pattern.
- Remediation should cover the whole Moni site with respect to As under all considered exposure scenarios, and Sb and Mn (except for the sub-area represented by one point) under the residential scenario and most part of the site with respect to Cr (VI) under the residential scenario.
- The HRA results, as well as the RBRLs, differ significantly among the considered exposure scenarios. Therefore, while generating remedial options, technical

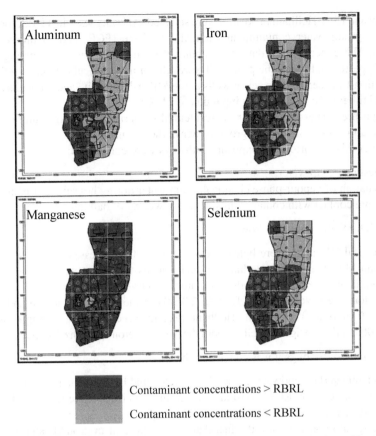

Contaminant concentrations > RBRL

Contaminant concentrations < RBRL

Fig. 5.8 Remedial zones for aluminium, iron, manganese and selenium under the residential scenario for the Moni site, Cyprus

feasibility and costs of remediation actions will depend on the decision regarding the future end land use.

- The established RBRLs values and remedial zones indicate that the remedial costs would be the lowest for the recreational scenario under the assumed exposure conditions, and would be lower for the industrial than the residential scenario.

5.2.2   Case Study 2 – Incore Project – Milan-Meton Site, Italy and Linz-Heilham, Austria

The INCORE project provided a strategy for the investigation, assessment and revitalisation of contaminated groundwater in urban industrialised areas. One of the project objectives was to assess health risk due to groundwater contamination by employing a site-specific risk assessment method [5]. To illustrate the application of this method to contaminated groundwater two INCORE test sites, *i.e.*, Milan-Meton in Italy and Linz-Heilham in Austria were selected. At both these sites HRA focused on two groundwater contaminants: tetrachloroethene (PCE) and trichloroethene (TCE). The concentration data were obtained from the immission pumping tests (IPTs), performed under the project. The results of a numerical evaluation of concentration-time series from seven wells at the Milan-Meton site and ten wells at the Linz-Heilham site were used as input data for HRA.

Exposure was assessed under the residential scenario for potential future receptors (adult residents), who use groundwater in their households.

Under this scenario three exposure pathways were considered:

- ingestion of tap water,
- inhalation of vapour-phase chemicals from water during household use, and
- dermal contact with contaminants while showering or bathing.

The HRA results were as follows:

- The total HI values were below the value of 1 in all investigated wells at both Milan-Meton and Linz-Heilham sites. It means that there was no concern of potential toxic effects under the residential scenario.
- The total CRs were from 5.0E-05 to 2.0E-04 at the Milan-Meton site and from 1.4E-06 to 6.2E-05 at the Linz-Heilham site, and exceeded the cancer target risk of 1E-06 in all investigated wells at both the Milan-Meton and Linz-Heilham sites.

Based on these findings it was concluded that:

- Contaminated groundwater at the Milan-Meton and Linz-Heilham sites may pose a potential cancer risk to consumers of groundwater under the residential exposure scenario.
- Taking into consideration the fact that TCE is present in groundwater as a byproduct or breakdown product of PCE, it would be advisable to identify and remove the sources of PCE pollution to reduce health risk.

5.3   Risk-Based Approach Role in Decision Making

The risk-based approach proposed for assessment of contaminated land in Cyprus and Poland constitutes, at the same time, a tool for supporting the remediation and decision-making process. Specifically, application of the risk-based approach to contaminated land assessment and remediation allows:

- to determine the needs for reducing risk and the needs for remedial/corrective actions,
- to set up remedial/corrective options at the site,
- to select a remediation option or a corrective action, appropriate for site-specific conditions,
- to design and conduct site remedial actions to protect human health, and
- to facilitate other risk management decisions (e.g., changing the land use pattern).

Acknowledgements The Moni data were generated during the project of the *"National Inventory of Potential Sources of Soil Contamination in Cyprus"* (Tender number 5/2004), and the results are published by permission of the Director of the Cyprus Geological Survey Department, and the General Director of the Hellenic Institute of Geology and Mineral Exploration.

References

1. Bardos P, Lewis A, Nortcliff S, Matiotti C, Marot F, Sullivan T (2003) CLARINET report "Review of decision support tools for contaminated land and their use in Europe". Austrian Federal Environment Agency, on behalf of CLARINET, Vienna
2. Carlon C (ed) (2007) Derivation methods of soil screening values in Europe. A review and evaluation of national procedures towards harmonization. Annex 2 – Country reports, EUR 22805-EN. European Commission, Joint Research Centre, Ispra
3. Ferguson C, Darmendrail D, Freier K, Jensen BK, Jensen J, Kasamas H, Urzelai A, Vegter J (eds) (1998) Risk assessment for contaminated sites in Europe, vol 1, Scientific basis. LQM Press, Nottingham
4. Ferguson CC, Kasamas H (eds) (1999) Risk assessment for contaminated sites in Europe, vol 2, Policy frameworks. LQM Press, Nottingham
5. Gzyl J, Wcislo E, Biesiada M, Gzyl G, Krupanek J (2004) Health risk assessment due to groundwater contamination. Ingegneria e Geologia degli Acquiferi (IGEA), special issue on the INCORE project (19): 101–108 (Supplemento a GEAM, A.XLI n.3)
6. Kuperberg JM, Wcislo E, Teaf CM (1996) Application of risk-based approaches to the remediation of a refinery site in Poland. In: Program of third international symposium and exhibition on environmental contamination in Central and Eastern Europe, Warsaw, 10–13 Sept 1996 (abstract p 148)
7. Sanka M (2007) Czech Republic. In: Carlon C (ed) Derivation methods of soil screening values in Europe. A review and evaluation of national procedures towards harmonization. Annex 2 – Country reports, EUR 22805-EN. European Commission, Joint Research Centre, Ispra, pp 145–149
8. USEPA (US Environmental Protection Agency) (1989) Risk assessment guidance for Superfund, vol I, Human health evaluation manual. Part A, Interim Final, EPA/540/1-89/002. Office of Emergency and Remedial Response, Washington, DC
9. USEPA (US Environmental Protection Agency) (1991a) Risk assessment guidance for Superfund, vol I, Human health evaluation manual, Part B, Development of risk-based preliminary remediation goals, Interim, EPA/540R-92/003, Publication 9285.7-01B. Office of Emergency and Remedial Response, Washington, DC
10. USEPA (US Environmental Protection Agency) (1991b) Role of the baseline risk assessment in Superfund remedy selection decisions, OSWER directive 9355.0–30. Office of Solid Waste and Emergency Response, Washington, DC
11. USEPA (US Environmental Protection Agency) (1991c) Risk assessment guidance for Superfund, vol I, Human health evaluation manual, Part C, Risk evaluation of remedial alternatives, Interim, Publication 9285.7-01C. Office of Emergency and Remedial Response, Washington, DC
12. USEPA (US Environmental Protection Agency) (1994) Memorandum: OSWER directive: revised interim soil lead guidance for CERCLA sites and RCRA corrective action facilities, EPA OSWER directive 9355.4-12. Office of Emergency and Remediation Response, Washington, DC
13. USEPA (US Environmental Protection Agency) (1996a) Soil screening guidance: technical background document, EPA/540/R-95/128, Publication 9355.4-17A. Office of Emergency and Remedial Response, Washington, DC
14. USEPA (US Environmental Protection Agency) (1996b) Soil screening guidance: user's guide, EPA/540/R-96/018, Publication 9355.4-23. Office of Emergency and Remedial Response, Washington, DC

15. USEPA (US Environmental Protection Agency) (1998) Memorandum: OSWER directive: clarification to the 1994 revised interim soil Lead (Pb) guidance for CERCLA sites and RCRA corrective action facilities. EPA/540/F-98/030. OSWER directive #9200.4-27P. Office of Solid Waste and Emergency Response, Washington, DC

16. USEPA (US Environmental Protection Agency) (2000) Superfund supplemental guidance to RAGS: region 4 bulletins, human health risk assessment bulletins, USEPA region 4, originally published November 1995. (http://www.epa.gov/region4/waste/ots/healtbul.htm - hhremed)

17. USEPA (US Environmental Protection Agency) (2001) Risk assessment guidance for Superfund, vol I, Human health evaluation manual, Part D, Standardized planning, reporting, and review of Superfund risk assessment, final, Publication 9285.7-47. Office of Emergency and Remedial Response, Washington, DC

18. USEPA (US Environmental Protection Agency) (2002) Supplemental guidance for developing soil screening levels for Superfund sites, OSWER 9355.4-24. Office of Solid Waste and Emergency Response, Washington, DC

19. USEPA (US Environmental Protection Agency) (2004) Risk assessment guidance for Superfund, vol I, Human health evaluation manual, Part E, supplemental guidance for dermal risk assessment, final, EPA/540/R/99/005, OSWER 9285.7-02EP, PB 99–963312. Office of Superfund Remediation and Technology Innovation, Washington, DC

20. USEPA (US Environmental Protection Agency) (2006) Frequent questions from risk assessors on the Adult Lead Methodology (ALM), (http://www.epa.gov/superfund/health/contaminants/lead/almfaq.htm - screening)

21. USEPA (US Environmental Protection Agency) (2011) Waste and cleanup risk assessment, (http://www.epa.gov/oswer/riskassessment/index.htm).

22. Wcislo E (2002) Risk-based approach to remediation of contaminated sites in Poland. In: Terytze K, Atanassov I (eds) Proceedings of the international workshop: assessment of the quality of contaminated soils and sites in Central and Eastern European Countries (CEEC) and New Independent States (NIS), Sofia, 30 Sept–3 Oct 2001. GortexPress, Sofia, pp 67–70

23. Wcislo E (2006) Examples of health risk assessment application for contaminated sites in the Upper Silesia, Poland. Report of the pilot study meeting – prevention and remediation in selected industrial sectors. Small sites in urban areas, Athens, 5–7 June 2006. Report No. 277 under the auspices of the North Atlantic Treaty Organization's Committee on the Challenges on Modern Society (NATO/CCMS), EPA 542-R-06-003 (abstract p 14)

24. Wcisło E (2009) Ocena ryzyka zdrowotnego w procesie remediacji terenów zdegradowanych chemicznie – procedury i znaczenie (Human health risk assessment in remediation process of contaminated sites – role and procedures), (in Polish). Wydawnictwo Ekonomia i Środowisko, Białystok

25. Wcislo E, Dlugosz J, Korcz M (2003) NORISC. Human health risk assessment framework for decision-making, Deliverable 20. Institute for Ecology of Industrial Areas, Katowice. (http://www.norisc.com/)

26. Wcislo E, Ioven D, Kucharski R, Szdzuj J (2002) Human health assessment case study: an abandoned metal smelter site in Poland. Chemosphere 47:507–515

27. Wcislo E, Korcz M (2006) Guidance on contaminated site risk assessment in Cyprus, 1. Site-specific human health risk assessment (HRA), 2. Risk-based guideline values for soil contaminants (RBSGVs), Part 5, In: National Inventory of Potential Sources of Soil Contamination in Cyprus. Institute of Geology and Mineral Exploration, Athens, Institute for Ecology of Industrial Areas, Katowice, GeoInvest Ltd., Nicosia

28. Wcisło E, Korcz M (2008) Zasady oceny ryzyka zdrowotnego na terenach zanieczyszczonych na Cyprze (Guidelines for human health risk assessment in contaminated sites in Cyprus). In: Proceedings of the IX conference of the Polish Society of Toxicology, Szczyrk, 8–12 Sept 2008, Acta Toxicol 16 (suppl.):13–14 (in Polish)

29. Wcislo E, Korcz M, Dlugosz J, Owczarska I (2007) Human health risk assessment for the Moni Industrial Area, Cyprus. Part 10. In: National Inventory of Potential Sources of Soil Contamination in Cyprus, Institute of Geology and Mineral Exploration, Athens, Institute for Ecology of Industrial Areas, Katowice, GeoInvest Ltd., Nicosia

Chapter 6
Comparative Measurements of Radon Content in Tap Water in Tbilisi and Rustavi Cities

Nodar P. Kekelidze, Gia Kajaia, Teimuraz V. Jakhutashvili, Eremia V. Tulashvili, Lela A. Mtsariashvili, and Zaur Berishvili

Abstract Results are presented of long-term periodic measurements of content of radioactive gas – radon ^{222}Rn in drinking (tap) water from various districts in Tbilisi as a function of time during years 2009 and 2010. It is shown, that according to the level of radon content in tap water, the investigated districts can be divided conditionally into two groups: with rather high radon content – more than 1 Bq l^{-1} (Vake, Saburtalo, Ortachala), and districts with rather low radon content – less than 1 Bq l^{-1} (Digomi, Isani, Vazisubani). It is underlined that this circumstance can be connected with presence of two qualitatively distinguished sources of tap water supply in Tbilisi – underground (artesian wells near settlements Natakhtari, Bulachauri, Mukhrani) and surface (the Tbilisi reservoir filled with waters of the rivers Aragvi and Iori) in which conditions of aeration and degassing by radon have an essentially different character. Also some other features in the character of radon distribution are marked. Results of radon content researches in tap water in another large city in Georgia – Rustavi – are also given. Radon concentration varied in the range 0.06–16.5 Bq l^{-1}. Average values of radiological parameters have been estimated: in Tbilisi – committed effective dose ($7.7 \cdot 10^{-5}$ to $1.1 \cdot 10^{-2}$ mSv·year^{-1}) and dose equivalent to the stomach ($2.2 \cdot 10^{-3}$ to $3.3 \cdot 10^{-1}$ mSv·year^{-1}), and in Rustavi – committed effective dose ($1.5 \cdot 10^{-4}$ to $4.2 \cdot 10^{-2}$ mSv·year^{-1}) and dose equivalent to the stomach ($4.4 \cdot 10^{-3}$ to $1.2 \cdot$ mSv·year^{-1}).

N.P. Kekelidze (✉) • G. Kajaia • T.V. Jakhutashvili • E.V. Tulashvili • L.A. Mtsariashvili
• Z. Berishvili
Material Research Institute, Faculty of Exact and Natural Sciences, Iv. Javakhishvili
Tbilisi State University, 3, Av. Chavchavadze, Tbilisi 0128, Georgia
e-mail: n.kekelidze@tsu.ge

F.F. Quercia and D. Vidojevic (eds.), *Clean Soil and Safe Water*, NATO Science for Peace
and Security Series C: Environmental Security, DOI 10.1007/978-94-007-2240-8_6,
© Springer Science+Business Media B.V. 2012

6.1   Introduction

Water has a significant place in human life. One of the major problems of research of surface waters properties, including drinking water, is the presence of certain amounts of radioactive gas – radon (^{222}Rn).

Radon content studies in drinking water, in particular in tap water, is the subject of numerous researches carried out in many countries in the world during last decades. As an example, one study [6] informs that radon content in tap water in some cities and settlements located on the east coast of the Black Sea in Turkey changes in the range 5.31–18.46 Bq l^{-1}; radon concentration in drinking water of municipal water supply in Lahore, Pakistan ranges from 2.0 up to 7.9 Bq l^{-1} [13], and in the drinking water municipal supply system r in the city of Baoji, China ranges in the interval 8–18 Bq l^{-1} [23], etc.

The radon problem in Georgia was not given proper attention in the past. Practically, until recently, there were no data of radon content in drinking (tap) water. Some results of radon content researches in natural water and soil were given in a recent work [4], in particular in Vani, Chokhatauri, Ozurgeti regions (16–22 Bq l^{-1}). It is pointed out that areas with abnormal high radon exhalation in water have been revealed (however measured values were not reported).

In one more recent study [14, 15] data of radon content in drinking (including tap) water in Tbilisi – the capital of Georgia – have been reported.

The study of radon presence in various regions of Georgia and in environmental media, first of all, in water resources, is an actual task. In the present work results of researches of radon content in tap water in Tbilisi and in Rustavi – industrial centre in Georgia – are reported.

6.2   Investigated Media and Methodology

6.2.1   Sampling in Tblisi

Object of measurements were samples (465 samples in total) of the following types:

- drinking (tap) water – 460 samples selected in various districts in Tbilisi (Fig. 6.1) – Digomi (Dgm), Saburtalo (Sbrt), Vake (Vake), Ortachala (Ortch), Isani (Isani), Vazisubani (Vzsb);
- water samples from surroundings of Tbilisi – Natakhtari and Tbilisi reservoir (5 samples in total).

Sampling has been carried out, generally on a weekly basis, during years 2009 and 2010.

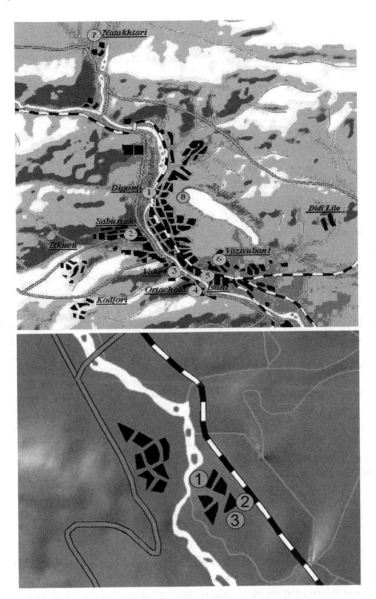

Fig. 6.1 Scheme of control points location in the territory of Tbilisi (*top*) and Rustavi (*bottom*). Tbilisi: *1* Digomi; *2* Saburtalo; *3* Vake; *4* Ortachala; *5* Isani; *6* Vazisubani; *7* Natakhtari; *8* Tbilisi reservoir; Rustavi: *1* Rst-1; *2* Rst-2; *3* Rst-3

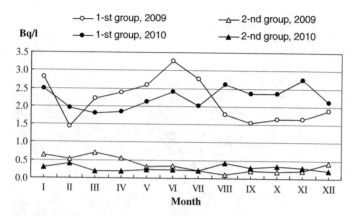

Fig. 6.2 Monthly average radon concentration in tap water depending on sampling date in years 2009 and 2010 for two groups of districts of Tbilisi

6.2.2 Sampling in Rustavi

Investigated media were tap water, sampled from the water supply system of Rustavi in three control points – Rst-1, Rst-2, Rst-3 (Fig. 6.2). Sampling has been carried out, generally on a monthly basis, during years 2009 and 2010. In total 30 samples were collected, in particular:

• 22 samples (including 6 analytical duplicates) at a filling station near a bridge across river Kura (control point Rst-1);
• 5 samples from the grocery store near the former Research-and-Production Association "Azot" (control point Rst-2);
• 3 samples from the canteen near the former Research-and-Production Association "Azot" (control point Rst-3).

6.2.3 Sampling Methodology

Water samples for radon measurement were collected in special containers (volume 250 ml). After water was filled up to the top, the container was hermetically closed by the cover, overturned up by a bottom, and transported to the laboratory.

6.2.4 Measurements

Measurement of radon activity concentration in water was carried out by alpha-spectrometry using Electronic Radon Detector RAD7 [11]. Measurements were carried out within several hours after sampling.

6.2.5   Estimation of Doses

Estimate of radon doses by ingestion was made by use of conversion factors widely accepted in the literature [10]; in particular, the following parameters were estimated:

$$\text{committed effective dose D(ef), mSv·year}^{-1} : D(ef) = A \cdot C_1 \cdot Y$$

where:

A – radon concentration in water, $Bq \cdot l^{-1}$,
C_1 – conversion factor, equal to $0.35 \cdot 10^{-8}$ Sv Bq^{-1} [21],
Y – annual consumption of 730 l was assumed based on a conservative consumption of 2 l day^{-1} per year for a "standard adult" [18, 20];
dose equivalent to the stomach D(eq), mSv·year^{-1} (as human organ most sensitive to radiation at consumption of water): $D(eq) = A \cdot C_2 \cdot Y$
where:

C_2 – conversion factor, equal to $1 \cdot 10^{-7}$ Sv Bq^{-1} [7].

6.3   Experimental Results

6.3.1   Tbilisi

In Table 6.1 average, minimal and maximal values of monthly average radon concentration depending on sampling points for various districts in Tbilisi during years 2009 and 2010 are given.

Results of measurements of radon concentration in several samples of tap water collected in the settlement of Natakhtari (near the corresponding source of tap water supply) and in one water sample of Tbilisi reservoir are given in Table 6.2.

Apparently, from the data shown in Tables 6.1 and 6.2, all districts can be divided into two groups by the radon content:

- 1st group: districts with rather high radon content, more than 1 Bq l^{-1} (Vake, Saburtalo, Ortachala);
- 2nd group: districts with rather low radon content, less than 1 Bq l^{-1} (Digomi, Isani, Vazisubani);
- among the districts of 1st group it is possible to highlight district Vake, where the average radon content – 3.4 and 2.7 Bq l^{-1} for 2009 and 2010; which is appreciably higher – approximately 1.1–2.8 times, than in Saburtalo and Ortachala districts;
- among districts of 2nd group the radon average content in Vazisubani is approximately 2 times smaller than in districts Digomi and Isani;
- values of radon concentration in settlement Natakhtari correspond to the values of 1st group of districts, and in Tbilisi reservoir to the values of 2nd group of districts.

Table 6.1 Average (ave), minimal (min) and maximal (max) values of monthly average radon concentration (Bq l⁻¹) at different sampling points in various districts of Tbilisi during years 2009 and 2010

Parameters	Dgm	Sbrt	Vake	Ortch	Isani	Vzsb
2009						
ave	0.38	1.8	3.4	1.2	0.53	0.17
min	0.20	0.82	2.5	0.09	0.03	0.06
max	0.67	2.4	4.5	3.0	1.7	0.33
2010						
ave	0.39	1.8	2.7	2.4	0.24	0.17
min	0.27	1.4	2.0	2.1	0.05	0.09
max	0.49	2.2	3.7	2.7	0.6	0.28

Table 6.2 Results of measurements of radon concentration (A, Bq l⁻¹) in samples of tap water in Natakhtari and water of Tbilisi reservoir

Area	Month, year	A, Bq l⁻¹
Natakhtari	III-09	4.1
-"-	V-09	3.5
-"-	IX-09	4.6
-"-	XI-09	3.8
Tbilisi reservoir	VII-09	0.14

Monthly average values of radon concentration in tap water samples in years 2009 and 2010 for two groups of districts are given in Fig. 6.2.

Apparently from the graph it is possible to note for the 1st group of districts an increase of radon concentration in winter and summer months and a certain decrease in winter and autumn months. Such temporary dependence is not observed for the 2nd group.

Table 6.3 shows average, minimal and maximal values of radon concentration in investigated tap water samples at various sampling sites and estimated values of the committed effective dose and dose equivalent to the stomach.

Apparently from Table 6.3 values of the committed effective dose for Tbilisi ranged from $7.7 \cdot 10^{-5}$ to $1.1 \cdot 10^{-2}$ mSv year⁻¹, and values of dose equivalent to stomach ranged from $2.2 \cdot 10^{-3}$ to $3.3 \cdot 10^{-1}$ mSv year⁻¹.

6.3.2 Rustavi

The generalized results for years 2009 and 2010 – average, minimal and maximal values – of radon activity concentration (A) in tap water at control points are given in Table 6.4. In the same table corresponding values of the committed effective dose D(ef) and dose equivalent to stomach D(eq) are given. Figure 6.3 shows time dependence of radon content in water at control point Rst-1 where the largest values have been detected.

From Table 6.4 and Fig. 6.3, it is possible to note the following features:

- the largest value of average radon activity concentration (13.0 Bq l⁻¹) is observed in control point Rst-1, while the lowest in control point Rst-3 (0.08 Bq l⁻¹);

Table 6.3 Average (ave), minimal (min) and maximal (max) values of radon concentration (A) in investigated tap water samples at various sampling sites and estimated values of the committed effective dose D(ef) and dose equivalent to stomach D(eq) in various districts of Tbilisi, as well as in Natakhtari and Tbilisi reservoir

District	A, Bq l^{-1}			D(ef), mSv year^{-1}			D(eq), mSv year^{-1}		
	ave	min	max	ave	min	max	ave	min	max
2009									
1 group	2.1	0.09	4.5	$5.5 \cdot 10^{-3}$	$2.3 \cdot 10^{-4}$	$1.1 \cdot 10^{-2}$	$1.6 \cdot 10^{-1}$	$6.6 \cdot 10^{-3}$	$3.3 \cdot 10^{-1}$
2 group	0.36	0.03	1.7	$9.2 \cdot 10^{-4}$	$7.7 \cdot 10^{-5}$	$4.2 \cdot 10^{-3}$	$2.6 \cdot 10^{-2}$	$2.2 \cdot 10^{-3}$	$1.2 \cdot 10^{-1}$
Natakhtari	4.0	3.5	4.6	$1.0 \cdot 10^{-2}$	$8.9 \cdot 10^{-3}$	$1.2 \cdot 10^{-2}$	$2.9 \cdot 10^{-1}$	$2.5 \cdot 10^{-1}$	$3.4 \cdot 10^{-1}$
Tbilisi reservoir	0.14			$3.6 \cdot 10^{-4}$			$1.0 \cdot 10^{-2}$		
2010									
1 group	2.3	1.4	3.7	$5.8 \cdot 10^{-3}$	$3.6 \cdot 10^{-3}$	$9.4 \cdot 10^{-3}$	$1.7 \cdot 10^{-1}$	$1.0 \cdot 10^{-1}$	$2.7 \cdot 10^{-1}$
2 group	0.27	0.05	0.61	$7.0 \cdot 10^{-4}$	$1.2 \cdot 10^{-4}$	$1.6 \cdot 10^{-3}$	$2.0 \cdot 10^{-2}$	$3.5 \cdot 10^{-3}$	$4.4 \cdot 10^{-2}$

Table 6.4 Average (*ave*), minimal (*min*) and maximal (*max*) values of radon concentration (A) and estimated values of committed effective dose D(ef) and dose equivalent to the stomach D(eq) in investigated tap water samples at various control points of Rustavi in years 2009 and 2010

Control point	A, Bq l^{-1}			D(ef), mSv year^{-1}			D(eq), mSv year^{-1}		
	ave	min	max	ave	min	max	ave	min	max
Rst-1	13.0	8.9	16.5	$3.3 \cdot 10^{-2}$	$2.3 \cdot 10^{-2}$	$4.2 \cdot 10^{-2}$	$9.5 \cdot 10^{-1}$	$6.5 \cdot 10^{-1}$	1.2
Rst-2	11.3	5.3	14.8	$2.9 \cdot 10^{-2}$	$1.3 \cdot 10^{-2}$	$3.8 \cdot 10^{-2}$	$8.2 \cdot 10^{-1}$	$3.9 \cdot 10^{-1}$	1.1
Rst-3	0.08	0.06	0.11	$2.1 \cdot 10^{-4}$	$1.5 \cdot 10^{-4}$	$2.9 \cdot 10^{-4}$	$5.9 \cdot 10^{-3}$	$4.4 \cdot 10^{-3}$	$8.4 \cdot 10^{-3}$

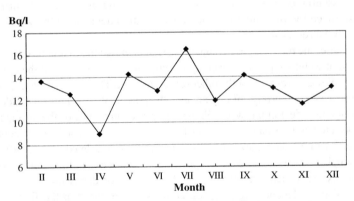

Fig. 6.3 Generalized monthly dependence of radon concentration in tap water at control point Rst-1 in years 2009 and 2010

- average radon activity concentration in control point Rst-2 (11.3 Bq l^{-1}) is lower (about 13%) than corresponding value for point Rst-1 while in control point Rst-3 it is approximately 163 times lower;

- during an annual cycle the tendency to increase of radon activity concentration in the autumn-winter period and to reduce in spring, with a local rise in May is observed;
- values of radiological parameters lie in the following ranges:
  - committed effective dose D(ef): $1.5 \cdot 10^{-4}$ to $4.2 \cdot 10^{-2}$ mSv year^{-1};
  - dose equivalent to the stomach D(eq): $4.4 \cdot 10^{-3}$ to 1.2 mSv year^{-1}.

6.4   Analysis

6.4.1   Tbilisi

Water supply to Tbilisi is provided basically by two qualitatively different sources – groundwater and surface water.

Groundwater sources are waters of the Aragvi gorge from artesian wells near settlements Natakhtari, Bulachauri, and Mukhrani. As surface source waters from the Tbilisi reservoir, which replenishes with water of the rivers Aragvi and Iori, are used. Thus, there are evidences to consider that formation conditions of radon composition of these sources of tap water essentially differ from each other. Apparently, water from artesian wells has an increased radon content (as shown from increased values of radon contents in the settlement Natakhtari, located in immediate proximity to the Natakhtari water supply system). In the water supply system fed by surface waters an essential degassing takes place before entering the system of purification plants, as proved by low values of radon concentration detected in the water sample from the Tbilisi reservoir.

Further, after the relevant processing, these waters merge into a unique city distribution system of water supply. However, apparently, in certain districts of the city, original features of water from one or the other source are kept. This is a possible explanation of the presence of two groups of city areas with appreciably distinguished radon contents (about one order) in tap water. Some differences in values of the average radon content detected within each group (such as a slightly increased radon content in Digomi and Isani in comparison to Vazisubani, etc.) can be related to local differences in water supply systems in these areas – presence or absence of buffer pools, reservoirs, pump stations, etc. which can influence the degree of radon degassing.

Observed time dependence of radon content in tap water in the first group of districts can be caused by annual variations of radon content in soil gas (that may affect concentration of dissolved radon in Natakhtari artesian sources); however also an influence of features in change of speed of aeration and degassing of radon from water in water supply systems is considered possible. The study of features of seasonal dependence of radon content and also the study of distribution of radon content in nearby soils is of certain scientific interest, and a dedicated research in this direction represents an actual need.

As shown from the obtained results, all values of radon concentration in tap water in various districts of Tbilisi are below the generally accepted maximum-permissible value (from 2 up to 100 times below). Average values of the committed effective dose and equivalent dose to the stomach ranged from 0.00092 to 0.0055 and from 0.026 to 0.16 mSv·year^{-1}, accordingly. These values correspond to data cited in literature, as for example, to data available from the city of Amasya, Turkey [16] and Tehran, Iran [3].

6.4.2 Rustavi

Relatively high values of radon concentration, frequently observed in tap water, can be related to the specificity of the water supply system in Rustavi City. Rustavi is supplied with water from artesian wells by means of two main sluices, originated at the junction of the two rivers Khrami and Debed. Thus the source of drinking water of the city is represented by groundwater which may contain a lot of radon. Its concentration directly on the output at the consumer, despite a rather lengthy sluices (about 20 km) and the presence of water-processing systems, still may remain enough high, as it is observed in the majority of samples. It is necessary to point out that in the city a certain degree of deficiency of drinking water is observed, and some consumers use local storage tanks in which water can be stored during the year. As a result an additional decrease of radon concentration takes place following decontamination and decay. This circumstance can cause significant decrease of the radon content level in some cases as, for example, in control point Rst-3 where low values of radon concentration were detected.

Some trends observed in present work in the dependence of radon content with time, in general, follow the seasonal fluctuations reported in the literature (for example, see [12]). However it is necessary to note that these trends can also result from interference of seasonal factors with duration of water processing technological procedures. This issue requires additional investigations.

In general resulted data show that the average radon content in tap water in Rustavi is appreciably higher than in Tbilisi. This circumstance can be explained by the greater radon saturation of the groundwater used in Rustavi, related to features of the geological and soil structure of this region. However, probably, also anthropogenic factors − such as shorter duration of the water cycle from source to consumer − can cause higher concentration in tap water of Rustavi.

Unfortunately up to now, there is no specific national regulation for radioactivity concentrations in drinking water in Georgia. Measured values of radon activity concentration in tap water of Tbilisi and Rustavi, basically, do not exceed maximal level of radon content in drinking water (11.1 Bq l^{-1}) recommended by the USEPA [9]. It is necessary to note that WHO established an higher level − 100 Bq l^{-1} − as normative value at which it is necessary to take appropriate measures on reduction of radon level in drinking water [22].

Table 6.5 Radon activity concentration (A), committed effective dose D(ef) and dose equivalent to the stomach D(eq) in tap water at different locations

Country, city	A, Bq l^{-1}	D(ef), mSv year^{-1}	D(eq), mSv year^{-1}	References
Kenya	0.8–4.7	–	–	Otwoma et al. (1998)
Egypt	0.07–2.33	–	–	[1]
Turkey	5.31–18.46	–	–	[6]
Jordan	2.5–4.7	–	–	[2]
Pakistan, Lahore	2.0–7.9	$7.3 \cdot 10^{-3}$ to $2.88 \cdot 10^{-2}$	–	[13]
China, Baoji	8.0–18.0	–	–	[23]
Iran, Mashhad	0.064–49.088	$1.2 \cdot 10^{-5}$ to $8.84 \cdot 10^{-3}$	–	[5]
Georgia, Ozurgeti	10.3 ± 2.75	0.0285 ± 0.0076	0.75 ± 0.37	[8]
Georgia, Tbilisi, 1st group of districts	0.09–4.5	$2.3 \cdot 10^{-4}$ to $1.1 \cdot 10^{-2}$	$6.6 \cdot 10^{-3}$ to $3.3 \cdot 10^{-1}$	Present work
Georgia, Tbilisi, 2nd group of districts	0.03–1.7	$7.7 \cdot 10^{-5}$ to $4.2 \cdot 10^{-3}$	$2.2 \cdot 10^{-3}$ to $1.2 \cdot 10^{-1}$	Present work
Georgia, Rustavi	0.06–16.5	$1.5 \cdot 10^{-4}$ to $4.2 \cdot 10^{-2}$	$4.4 \cdot 10^{-3}$ to 1.2	Present work

Comparison of measured data with some results from the literature is carried out in Table 6.5. Apparently, radon content in tap water of Tbilisi and Rustavi is comparable to radon content in tap water of some other investigated locations, in particular in Turkey, in Georgia (Ozurgeti), China (Baoji), and appreciably exceeds radon content measured in other countries, such as Kenya and Egypt.

Comparison of the estimated values of some radiological parameters with the data of other studies shows, that, for example, maximal values of D(ef) for tap water in Tbilisi (1st group of districts) are comparable with data from Mashhad, Iran, and are slightly greater than the data from Ozurgeti, Georgia, but are 3 times lower with respect to data from Lahore, Pakistan. Maximal values of D(ef) at control points Rst-1 and Rst-2 in Rustavi are approximately 1.5 times the values estimated for Ozurgeti (Georgia) and Lahore (it is necessary to note, that in the latter case authors used as value of conversion factor C_1 equal to $5 \cdot 10^{-9}$ Sv Bq^{-1}, which was recommended by UNSCEAR in 1988 [19]), and 5–7 times higher with respect to tap water in Mashhad, Iran and to 1st group of districts in Tbilisi. Maximal values of D(eq) at points Rst-1 and Rst-2 are also approximately 1.5 times higher than the values from Ozurgeti, Georgia and 5–7 times higher than values from tap water of 1st group of districts in Tbilisi.

6.5 Conclusions

1. Regular research studies have been carried out and quantitative results of radioactive gas – radon ^{222}Rn content in tap water in various districts of Tbilisi and Rustavi cities as a function if time (in years 2009 and 2010) have been obtained.

2. It is shown that, with respect to the level of radon content in tap water, the investigated districts of Tbilisi City can be divided into two groups:

   - with rather high radon content – more than 1 Bq l^{-1} (Vake, Saburtalo, Ortachala);
   - with rather low radon content – less than 1 Bq l^{-1} (Digomi, Isani, Vazisubani).
   - It is underlined that this circumstance can be connected with presence of two qualitatively distinguished sources of tap water supply in Tbilisi by – groundwater (artesian wells near settlements Natakhtari, Bulachauri, Mukhrani) and surface water (the Tbilisi reservoir filled with waters of the rivers Aragvi and Iori) for which conditions of aeration and radon degassing have an essentially different character.

3. Estimated values of radon concentration in tap water in Rustavi City range in the interval 0.06–16.5 Bq l^{-1}; it is shown, that values of activity of about 10 Bq l^{-1} are related to specific character of the source of water supply to the city – groundwater from artesian wells – in which radon content is apparently sufficiently high; in some cases much smaller values of activity (less than 1 Bq l^{-1}) – related to the use by consumers of storage tanks in which evaporation and decay takes place – are observed.

4. Parameters of radiological effects at water consumption points have been determined. In particular, in Tbilisi, average values of the committed effective dose ranged from $7.7 \cdot 10^{-5}$ up to $1.1 \cdot 10^{-2}$ mSv·year^{-1} at various control points, and the dose equivalent to the stomach ranged from $2.2 \cdot 10^{-3}$ up to $3.3 \cdot 10^{-1}$ mSv·year^{-1}. In Rustavi committed effective dose ranged from $1.5 \cdot 10^{-4}$ up to $4.2 \cdot 10^{-2}$ mSv·year^{-1} and the dose equivalent to the stomach ranged from $4.4 \cdot 10^{-3}$ up to 1.2 mSv·year^{-1}.

5. All measured values of radon content in tap water and estimates of radiological parameters lie generally below recommended thresholds of maximum-permissible concentration given in the literature.

Acknowledgments This work has been executed within the framework of project: GNSF/ PRES08/5-331 "Characteristic features of drinking water in Tbilisi area and its close regions".

References

1. Abbady A, Ahmed NK, Saied MH, El-Kamel AH, Ramadan S (1995) Variation of ^{222}Rn concentration in drinking water in Qena. Bull Fac Sci 24(1-A):101–106
2. Al-Bataina BA, Ismail AM, Kullab MK, Abmurad KM, Mustapha H (1997) Radon measurements in different types of natural waters in Jordan. Radiat Meas 28:1–6
3. Alirezazadeh N (2005) Radon concentration in public water supplies in Tehran and evaluation of radiation dose. Iran J Radiat Res 3(2):79–83
4. Amiranashvili AG, Chelidze TL, Gvinianidze KG, Melikadze GI, Todadze MSh, Trekov IY, Tsereteli DG (2008) Radon distribution and prevalence of lung cancer several areas of west Georgia. Trans Georgian Inst Hydrometeorol 115:349–353

5. Binesh A, Mohammadi S, Mowlavi AA, Parvaresh P, Arabshahi H (2010) Evaluation of the radiation dose from radon ingestion and inhalation in drinking water sources of Mashhad. Res J Appl Sci 5(3):221–225
6. Çevik U, Damla N, Karahan G, Çelebi N, Kobya I (2006) Natural radioactivity in tap waters of eastern Black Sea region of Turkey. Radiat Prot Dosim 118(1):88–92
7. de Oliveira J, Paci Mazzili B, de Oliveira Sampa MH, Bambalas E (2001) Natural radionuclides in drinking water supplies of Sao Paulo State, Brazil and consequent population doses. J Environ Radioact 53(1):99–109
8. Elizbarashvili M, Khikhadze N, Kekelidze N, Jakhutashvili T, Tulashvili E (2010) Radon content in sea, river, drinking water of resort zone Ureki-Shekvetili in west Georgia and estimation of radon-produced hazard for population. In: Proceedings of the 10th International Multidisciplinary Scientific Geo-Conference (SGEM 2010) – "Modern Management of Mine Producing, Geology and Environmental Protection", vol II, Albena Resort, Varna, Bulgaria, pp 47–54
9. Federal Register (1991) Part II. Environmental Protection Agency. 40 CFR Parts 141 and 142. National Primary Drinking Water Regulations; Radionuclides; Proposed Rule. (56 FR 33050) 18 July 1991
10. Galán López M, Martín Sánchez A, Gómez Escobar V (2004) Estimates of the dose due to ^{222}Rn concentrations in water. Radiat Prot Dosim 111(1):3–7
11. Jakhutashvili T, Tulashvili E, Khikhadze N, Mtsariashvili L (2009) Measurement of radon contents with use of modern mobile alpha-spectrometer RAD-7. Ga Chem J 9(2):179–182
12. Kogan RM, Nazarov IM, Fridman ShD (1976) Basics of environmental gamma-spectrometry. Atomizdat, Moscow
13. Manzoor F, Alaamer AS, Tahir SNA (2008) Exposures to 222Rn from consumption of underground municipal water supplies in Pakistan. Radiat Prot Dosim 130(3):392–396
14. Mtsariashvili L, Khikhadze N, Kekelidze N, Jakhutashvili T, Tulashvili E (2010) Radioactivity of tap water of Tbilisi City (Georgia) and estimation of radiological risk for population. In: Proceedings of the 10th International Multidisciplinary Scientific Geo-Conference (SGEM 2010) – "Modern Management of Mine Producing, Geology and Environmental Protection", vol II, Albena Resort, Varna, Bulgaria, pp 39–46
15. Mtsariashvili L, Khikhadze N, Tsintsadze M, Kekelidze N, Jakhutashvili T, Tulashvili E (2010) Natural radio-activity of drinking water in suburbs of Tbilisi. Ga Chem J 10(1):113–120
16. Oner F, Yalim HA, Akkurt A, Orbay M (2009) The measurements of radon concentrations in drinking water and the Yeşilirmak River water in the area of Amasya in Turkey. Radiat Prot Dosim 133(4):223–226
17. Otwoma D, Mustapha AO (1998) Measurement of ^{222}Rn concentration in Kenyan groundwater. Health Phys 74:91–95
18. Salonen L (1988) Natural radionuclides in groundwater in Finland. Radiat Prot Dosim 24(1–4):163–166
19. United Nations Scientific Committee on the Effects of Radiation (1988) Sources and effect of ionizing radiation. UN, New York
20. UNSCEAR (1993) Sources and effects of ionization radiation. Report to the General Assembly. UNSCEAR, New York
21. Wallström M (2001) Commission recommendation of 20 December 2001 on the protection of the public against exposure to radon in drinking water supplies. 2001/928/EURATOM (DOCE L 344/85 of 28/12/01, Brussels)
22. World Health Organization (1993) Guidelines for drinking water quality. WHO, Geneva
23. Xinwei L (2006) Analysis of radon concentration in drinking water in Baoji (China) and the associated health effects. Radiat Prot Dosim 121(4):452–455

Chapter 7
Manganese and Sulphate Background in Groundwater at Portoscuso (Sardinia): A Tool for Water Management in a Large Contaminated Area

Antonella Vecchio, Maurizio Guerra, and Gianfranco Mulas

Abstract The definition of background values in groundwater is an important issue for the assessment of groundwater contamination caused by local sources. The Italian Institute for Environmental Protection and Research (ISPRA) developed in 2009 the "Technical Guidelines for the evaluation of background levels of inorganic substances in groundwater". In these Guidelines the definition of the specific geological context, the evaluation of geochemical, chemical and biological phenomena determining the diffuse presence of chemicals in groundwater, the criteria for analysis of existing data, the minimum criteria for the geological and chemical characterization of the groundwater body, the procedures for spatial assessment, statistical analysis of groundwater data and finally the selection of the representative value for background have been set. This paper describes the application of ISPRA Technical Guidelines to the large area of Portoscuso Municipality (SW Sardinia) characterized by a diffuse passive contamination due to a metallurgic industrial district.

7.1 Introduction

The definition of background va lues in groundwater is an important issue for the assessment of groundwater contamination caused by local sources. Under the current Italian legislation on contaminated sites [6] two assessment steps are provided:

A. Vecchio (✉) • M. Guerra
Soil Protection Department, Institute for Environmental Protection and Research (ISPRA),
Via Vitaliano Brancati 48, Rome, Italy
e-mail: antonella.vecchio@isprambiente.it

G. Mulas
Municipality of Portoscuso, Via Marco Polo 1, Portoscuso, CI 09010, Italy

F.F. Quercia and D. Vidojevic (eds.), *Clean Soil and Safe Water*, NATO Science for Peace and Security Series C: Environmental Security, DOI 10.1007/978-94-007-2240-8_7, © Springer Science+Business Media B.V. 2012

1. tabular screening levels (Contamination Threshold Concentrations – CTCs) used for the identification of "potentially contaminated sites";
2. risk-based target levels (Risk Threshold Concentrations – RTC) as reference for the definition of "contaminated sites" and target clean-up levels.

Screening levels for soil and groundwater are tabulated in a specific Annex to the Decree (Annex V). For the majority of chemicals included in Annex V, the groundwater CTCs are coherent with drinking water standards (i.e. tap water standards). However, if a potentially contaminated site is located within an area where natural or anthropogenic (diffuse) phenomena caused exceeding of CTCs for soil and/or for groundwater, then screening levels are assumed equal to background values. These background values may account for the natural scenario as well as diffuse, anthropogenic contamination. Therefore background values in groundwater may also be determined for anthropogenic substances (typically organic contaminants).

The Italian Ministry for the Environment asked ISPRA (Institute for Environmental Protection and Research) to develop a methodology for the evaluation of background values in groundwater in the context of the assessment of large potentially contaminated areas, like Italian National Priority List Sites. ISPRA published in 2009 the "Technical Guidelines for the evaluation of background levels of inorganic substances in groundwater" [4] with the following limitations [10]:

- Guidelines can be applied to inorganics, metals and metalloids;
- the definition for Background Value of the Groundwater Directive [3] has been adopted (i.e. accounting basically natural geochemical contaminants inputs in groundwater);
- Guidelines are structured for the application in the context of contaminated sites identification and remediation and not for the definition of the "status of groundwater body".

The procedure proposed in the Guidelines values includes (see Fig. 7.1):

- the definition of the specific geological context;
- the evaluation of geochemical, chemical and biological phenomena determining the diffuse presence of chemicals in groundwater;
- the analysis of existing data (inclusion\exclusion criteria);
- the minimum requirements for planning new geological and chemical characterization of the groundwater body;
- the spatial assessment and statistical analysis of groundwater data
- the selection of the representative value for background;
- the "statistical approach" of population distribution data comparison for the background vs site data assessment.

This paper describes the application of ISPRA Technical Guidelines to the large area of Portoscuso Municipality (SW Sardinia) characterized by a diffuse, passive contamination due to the Portovesme metallurgic industrial district.

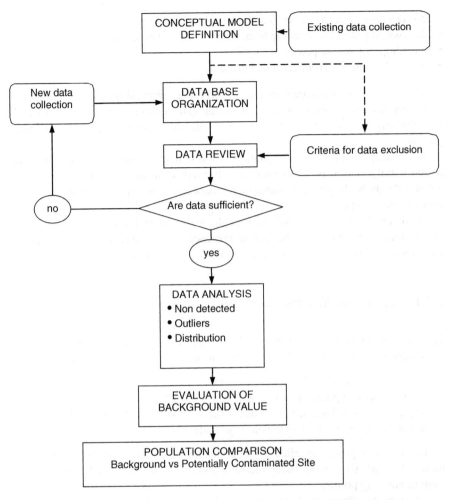

Fig. 7.1 Flow chart of the procedure for background values evaluation [10]

7.2 The Portoscuso Municipality Potentially Contaminated Area

The potentially contaminated area of Portoscuso Municipality (South-West Sardinia) has been included in the National Priority List Site "Sulcis Iglesiente Guspinese" for the presence of a big metallurgic industrial district. The "Portovesme industrial district" is characterized by the following potential polluting activities [1]:

- production of primary aluminum and related manufactures (laminates);
- production of non ferrous metals (sulfuric acid, copper cement, cadmium, metallic mercury, lead, sponge cadmium, copper spume, zinc);

- power production from combustion of coal and fuel oil;
- disposal of red muds resulting from bauxite refining process;
- metallurgic waste disposals;
- disposal of cinders from combustion plant.

The surrounding areas are principally used for agricultural purposes (production of vegetables and vineyards), but two residential agglomerates (Portoscuso and Paringiaunu) and a naturalistic valuable area (Boi-Cerbus Lagoon) are also present.

The conceptual model of soil contamination indicate a diffuse passive contamination mainly caused by emissions from the industrial district and the contaminated soil may be the primary source of groundwater contamination up gradient the Portovesme area [1].

However, the geological and hydrogeological context indicates also a natural geochemical contribution to the diffuse presence of inorganic substances (metals, metalloids and other inorganics) in soil and, to a minor extent, in groundwater.

7.2.1 Results of Site Investigation

The whole municipal territory outside the industrial area (30 km^2) has been investigated between July 2009 and March 2010. Investigations included (see Fig. 7.2):

- 62 surface probes (0–1.5 m below ground);
- 139 intermediate probes (from ground to capillary fringe);
- 66 piezometers: 40 surface piezometers (up to 15–25 m below ground level) and 26 deep ones (up to 40–133 m below ground);

The sampling procedure allowed the collection of 308 top soil (0–10 cm) samples, 371 surface (0–1 m) and deep soil samples (>1 m) and 78 groundwater samples from monitoring points.

Groundwater chemical investigation included, in addition to contaminants of interest, Physical/chemical parameters (temperature, redox potential, electric conductivity, pH, dissolved oxygen), total suspended soils and major ions (K^+, Na^+, Mg^{2+}, Ca^{2+}, Cl^-, SO_4^{2-}, HCO_3^-, NO_3^-) [2, 10].

Results of the investigation in the surrounding areas of the industrial district confirm a diffuse presence in soil of Zn, Pb and Cd all over the entire 30 km^2 wide investigated area. For these elements, the contamination pattern is characterized by an impressive trend decreasing along with sampling depth. This is indicative of a surface soil contamination mainly due to fall-out from the industrial district, while in deep soil the presence of contaminants may be correlated to natural background (Fig. 7.3).

The hydrogeological scenario is characterized by the presence of a not-continuous shallow aquifer hosted in Quaternary deposits (mainly sands) and of a deep one in the Cenozoic pyroclastic fractured rocks. The connection between the two aquifers is not clear, also influenced by local stratigraphy and time-dependent groundwater recharge [5].

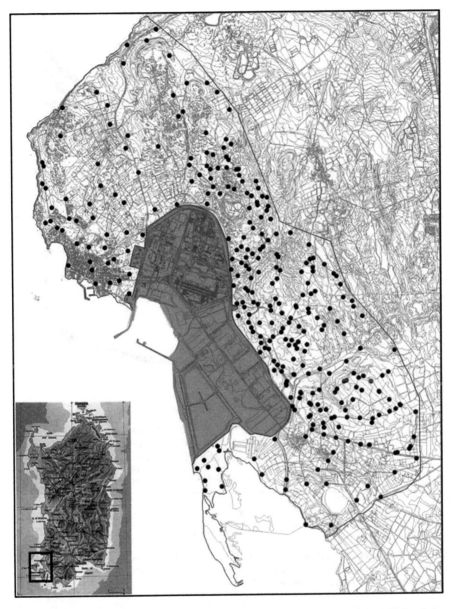

Fig. 7.2 The Portoscuso Municipality potentially contaminated area: sampling points of the investigation plan [5]

Inside the industrial district, groundwater within the shallow aquifer is characterized by a low oxidation level and reducing conditions, with an increase of pH and electric conductivity and a decrease of dissolved oxygen and redox potential downgradient [8].

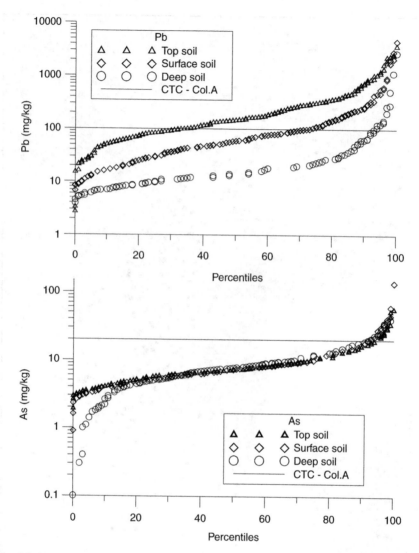

Fig. 7.3 Vertical contamination pattern of Pb. The significant decreasing of concentration with depth is emphasized. As reference, the depth-non sensitive pattern is reported for As, whose origin is not related to the fall out from industrial district

This situation is typical of shallow coastal groundwater with a low flow and limited recharge, but it can be amplified by the discharge of urban or industrial wastewaters with high organic content.

Concentrations of inorganic compounds (Pb, Cd, As, Al and F) in groundwater entering the industrial area are less than CTCs, but an enrichment in these elements is registered downgradient (see Fig. 7.4) with exceeding of screening values.

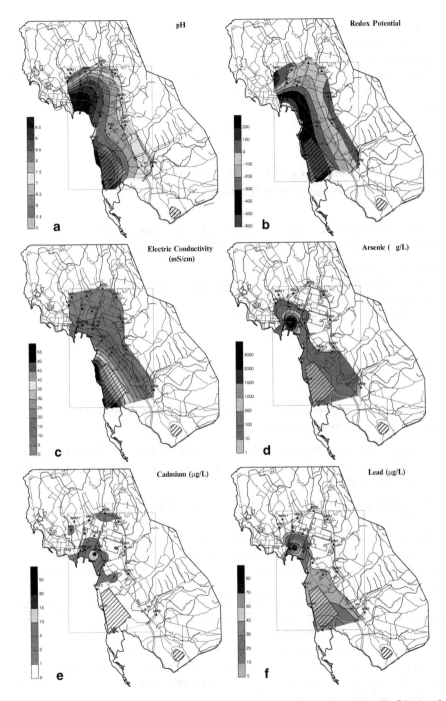

Fig. 7.4 Distribution of pH (**a**), redox potential (**b**), electric conductivity (**c**), As (**d**), Cd (**e**) and Pb (**f**) in groundwater inside the Portovesme industrial district [8]

Besides a diffuse presence in soil, Zn, Pb, Cd were found below screening values in the groundwater up gradient the industrial district; the contaminants exceeding thoroughly the CTCs were only Mn and SO_4.

Therefore for such chemicals the evaluation of groundwater background was required in order to identify potentially contaminated areas, according to the current legislation.

7.3 The Development of GW Background for Manganese and Sulphate

The assessment of groundwater background values at Portoscuso has been performed on the basis of 78 samples and included the following steps:

- data review (e.g. exclusion criteria);
- data analysis by means of statistical methods to describe the population of the data and to discriminate outliers, if any;
- calculation of a background range.

7.3.1 Data Review

To assess the suitability of collected data for the background evaluation the following "exclusion criteria" have been applied [4, 7]:

- presence of organic chemicals values more than 3 times of the Detection Limit;
- samples come from wells/piezometers whose screen position is not known (data may not belong to the analyzed aquifer).

The other two exclusion criteria reported in Muller et al. [7], have not been applied.

The exclusion of samples with NO_3 concentration more than 10 mg/l is considered too restrictive in the contest of management of contaminated sites, while this criterion is more appropriate for the assessment of "groundwater status" according to the Groundwater Directive.

The exclusion of samples with NaCl concentration more than 1,000 mg/l is not applicable for coastal aquifers, as in the case of Portoscuso. These criteria are going to be revalued also in ISPRA guidelines.

The hydrochemical characterization showed that groundwater of shallow aquifer in sands shows a general HCO_3-Ca pattern with some areas tending to a Cl-Na chemistry. Volcanic aquifer supports a more variable behavior taking into account mixing with shallow water (Fig. 7.5).

After the review of data, the suitable dataset included 56 records from 39 sampling points, given that the following samples were excluded:

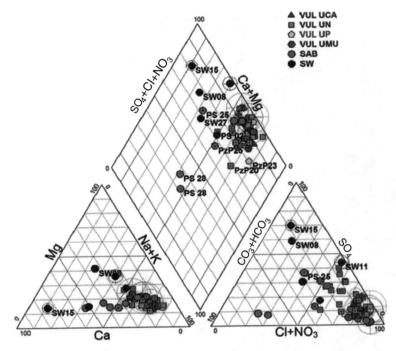

Fig. 7.5 Piper Diagram of Portoscuso groundwater. *VUL UCA, VUL UN, VUL UP, VUL UMU* indicates samples from volcanic aquifer, *SAB* indicates samples from quaternary sands deposits and *SW* indicates samples from the surroundings of "Waeltz roads" (roads paved with slags from industrial activities) [5]

- 5 records from 3 monitoring points with different hydrochemical facies with respect to the others;
- 5 records from surface monitoring points located just along the "Waeltz roads";
- 12 samples characterized by organic contamination.

7.3.2   Data Analysis

According to ISPRA Guidelines (Fig. 7.1) the analysis of data included [10]:

- "non-detected" evaluation: when data are recorded below the detection limit, to accomplish conservative assumption, the lowest available detection limit (DL) should be used as a datum;
- "outliers" identification: if there are "true outliers" (i.e. errors) in the dataset, they should be removed; "false outliers" (real extreme values) should be managed on the basis of site-specific conditions (contamination vs. geochemical anomaly);
- assessment of "data distribution": parametric and non-parametric tests application to select the appropriate probability data distribution (e.g. normal, lognormal, gamma, non parametric).

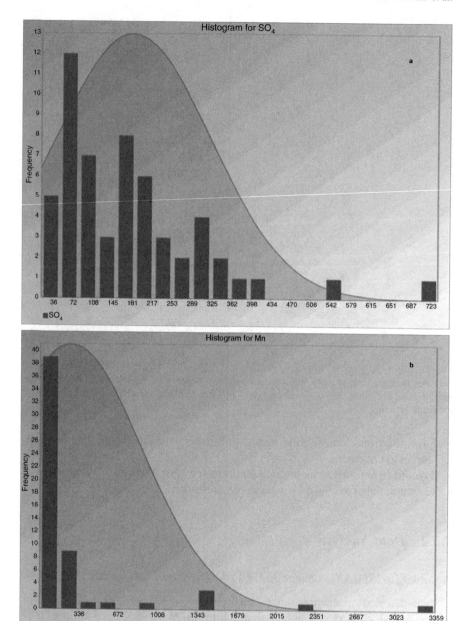

Fig. 7.6 Histogram of frequencies of the full background datasets for SO_4 (**a**) and Mn (**b**)

Table 7.1 Statistical indicators of background datasets with and without outliers for Sulphate and Manganese

Parameter	Unit	Sulphate Full dataset	Sulphate Without outliers	Manganese Full dataset	Manganese Without outliers
Number of observation	–	56	55	56	55
Distribution (95% confidence)	–	Gamma	Gamma	Lognormal	Lognormal
25 Percentile	[mg/L]	79	78	2	2
50 Percentile	[mg/L]	168	167	12	11
75 Percentile	[mg/L]	232	226	214	188
Mean	[mg/L]	182	172	264	208
Standard Deviation	[mg/L]	141	120	612	448
95 Percentile	[mg/L]	456	408	1,602	1,413
95 UPL (k=1)	[mg/L]	476	441	1,801	1,459
95 UTL/90%coverage	[mg/L]	456	423	1,486	1,218
95 UTL/95%coverage	[mg/L]	583	538	4,598	3,667

Statistical analysis have been performed using the ProUCL 4.00.05 software [9].

For Manganese values below the DL (8 over 56) were assumed equal to DL for conservative assumption. No SO_4 samples below the DL have been identified.

"True outliers" (sampling, analytical or reporting errors) have not been identified in the dataset.

Statistical tests for identification of potentially outliers have been applied and in particular the Rosner test (for n > 25, assuming normal distribution, up to 10 outliers) proposed by the ProUCL 4.00.05 software.

Sulphate shows a "Gamma distribution" with a 95% of confidence level (Fig. 7.6a). As full background dataset for SO_4 do not follow a normal distribution at 95% confidence level, a preliminary Gaussian transformation has been applied to data for performing Rosner test.

The comparison between the full dataset and that excluding the one potential outlier does not show significant differences in statistical indicators (Table 7.1).

Manganese shows a lognormal distribution with a 95% of confidence level (Fig. 7.6b). Therefore Rosner test has been performed on log-transformed data. The comparison between the full dataset and that excluding the single potential outlier shows more significant differences in statistical indicators than those found for Sulphate (Table 7.1).

7.3.3 Calculation of Background Range

For all datasets three statistical indicators of the right "tail" of the populations have been calculated and compared [9]:

- *95 percentile*: is the value of a variable below which a 95 percent of observations fall.

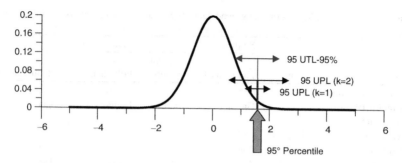

Fig. 7.7 Statistical indicators for the right "tail" of a population

- *95 Upper Tolerance Limit (UTL)*: defines the upper limit of the "Tolerance Interval" TI (limited below by the LTL). The Tolerance interval defines, at a certain confidence level, the range within falls a defined percentage, called "coverage", of the sampled population (Fig. 7.7).
- *95 Upper Prediction Limit (UPL)*: defines the upper limit of the "Prediction Interval" PI (limited below by the LPL). The Prediction Interval defines, at a certain confidence level, the range within we expect (i.e. with a probability more than 50%) that will fall a single observation (k = 1) or more observations simultaneously (k > 1) sampled from the same population of the dataset.

The relationships for this three statistical indicators are reported in Fig. 7.7.

These relationships are verified for the examined background dataset (Table 7.1). The general accordance between the statistical indicators, as in the case of sulphate dataset, indicates a good agreement of sampled dataset and population data in the "critical" region of the right tail. Greater differences have been found for Manganese dataset.

According to ISPRA Guidelines, the 95 percentile has been selected as representative value for background. Therefore background for Sulphate was set at 450 mg/L (CTC 250 mg/L), while Manganese BV was fixed at 1,600 mg/L (CTC 50 mg/L).

The 30-times CTC (i.e. tap water standards) background value of Mn was supported by the evidence of Mn-nodules in volcanic rocks reported during drilling operations.

$$95 \text{ UPL} \left(k = 1\right) = 50 \text{ UTL} / 90\%$$

$$95^\circ \text{ percentile} < 95 \text{UPL} \left(k = 1\right) < 95 \text{UTL} - 95\%$$

The groundwater potential contamination pattern, considering the background, is limited to restricted parts of the concerned area, while the exceeding CTC zone interested large portion of the municipality territory (Fig. 7.8).

The definition of background values as "new" screening levels for Portoscuso groundwater upgradient the Portovesme industrial district, according to current legislation, allowed to declare large portions of municipality as "not contaminated" areas.

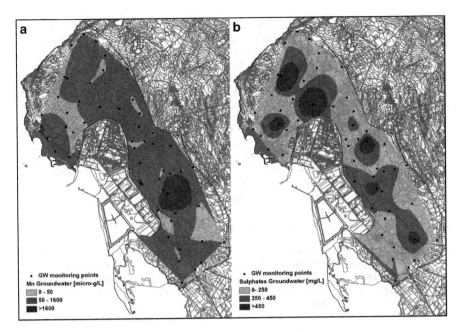

Fig. 7.8 Distribution of Mn (**a**) and SO$_4$ (**b**) in groundwater at Portoscuso

7.4   Conclusions

The definition of background values (BV) in groundwater is an important issue for the assessment of groundwater contamination caused by local sources.

Screening levels for soil and groundwater (CTCs) can be substituted by BV due to the natural scenario as well as diffuse, anthropogenic contamination.

The background levels obtained from this study allowed to declare large portion of Portoscuso Municipality as "non contaminated" areas, avoiding the application of useless an costly measures to groundwater outside the industrial district.

However in some cases samples exceeding the background values has been registered also in the upgradient municipal areas.

In these situations integrative investigations, including mineralogical analyses of soil materials and isotopic analyses, may be performed to exclude the presence of non-diffuse anthropogenic input to groundwater.

References

1. APAT (2008) Characterization plan of areas external to the Portovesme industrial district – Portoscuso Municipality (in Italian)
2. Chéry L (2006) Qualité naturelle des eaux souterraines. Méthode de caractérisation des états de référence des aquifères français, Guide technique. BRGM édition, 2006, 238 pp

3. European Union (2006) Directive 2006/118/EC of the European Parliament and of the Council of 12 December 2006 on the protection of groundwater against pollution and deterioration, Official Journal of the European Union of 27 December 2006, L 372/19

4. ISPRA (2009) Technical guidelines for the evaluation of background levels of inorganic substances in groundwater (in Italian). http://www.isprambiente.gov.it/site/it-IT/Temi/Siti_contaminati/Caratterizzazione_e_documentazione/

5. ISPRA, Portoscuso Municipality (2010) Characterization plan of areas external to the Portovesme industrial district – Results of investigations and risk assessment – Final Report (in Italian)

6. Italian Parliament (2006) Legislative Decree 3 April 2006, n. 152 – Environmental legislation, Official Gazette of the Italian Republic n.88 of 14 April 2006 – Ordinary Supplement no. 96

7. Muller D, Blum A, Hart A, Hookey J, Kunkel R, Scheidleder A, Tomlin C, Wendland F (2006) Final proposal for methodology to setup groundwater treshold values in Europe, Deliverable D18, BRIDGE project, 63 pp. www.wfd-bridge.net

8. RAS (2006) Remediation plan for the rehabilitation of Sulcis Iglesiente territory – Environment Quality Status (in Italian)

9. USEPA (2010) ProUCL version 4.00.05 technical guide, 235 pp. http://www.epa.gov/esd/tsc/software.htm

10. Vecchio A, Fratini M, Guerra M, Calace N (2010) Italian Guidelines for the evaluation of background values in groundwater, Report of the Joint NICOLE – COMMON FORUM Workshop: contaminated land management: opportunities, challenges and financial consequences of evolving legislation in Europe, pp 21–23. http://www.nicole.org/documents/DocumentList.aspx?l=1&w=n

Chapter 8
Contaminated Sites in Well Head Protection Areas: Methodology of Impact Assessment

Kestutis Kadunas

Abstract According to the Lithuanian legislation, protection zones of public water supply sources are divided into three strips on land: (1) "strip of strict protection", (2) "strip for protection against microbiological pollution" and (3) "strip for protection against chemical pollution". The strip of strict protection is the area immediately surrounding the abstraction point. It has to be of at least 5–50 m radius from the abstraction point (geometric criteria). The strip for protection against microbiological pollution is the territory surrounding the strip of strict protection. The size of the strip, depending on average annual abstraction with 15 years prediction, can be defined or calculated (computed). Application of protective measures on land use, in order to protect the resource from a qualitative and quantitative point of view, depends on natural protection level (vulnerability) of the primary aquifer. The strip for protection against chemical pollution is usually calculated or computed (taking in to account, that pollutant reached aquifer) using time criteria where potential pollutant travel time is 25 years (10,000 days). Two strips (strip 3a and strip 3b), with different protection measures applied, should be delimited. In 2010 more than 1,500 well fields (and single wells) were used to provide population with drinking water. For 650 well fields the sanitary protection strips were calculated or computed and required reports presented to Geological Survey. Totally sanitary protection zones cover about 1,300 km^2, which means about 2% of the territory of Lithuania. Inventory of more than 1,100 potential groundwater pollution sources, existing within sanitary protection strips, allowed assessing of possible impact to drinking water chemical composition. Assessment of possible or existing impact were carried out using indirect (no measurements) and direct (including measurements) methods. The results of this study are planned to be used for improvement of legislation regulating well head protection areas.

K. Kadunas (✉)
Geological Survey of Lithuania, S. Konarskio str. 35, Vilnius, LT 03123, Lithuania
e-mail: kestutis.kadunas@lgt.lt

F.F. Quercia and D. Vidojevic (eds.), *Clean Soil and Safe Water*, NATO Science for Peace and Security Series C: Environmental Security, DOI 10.1007/978-94-007-2240-8_8,

8.1   Drinking Water Protection Zone and Criteria for Delimitation

According to the Lithuanian legislation, protection zone of public water supply source are divided into three strips on land:

1. "strip of strict protection",
2. "strip for protection against microbiological pollution" and
3. "strip for protection against chemical pollution".

The *strip of strict protection (1 strip)* is the area immediately surrounding the abstraction point. It has to be of at least 5–50 m radius from the abstraction point (geometric criteria). The differences in diameter of the strip depend on the depth (vulnerability) to primary aquifer. Strip of strict protection must be fenced, and allowed activities are only related to water abstraction, treatment and other related infrastructures.

The *strip for protection against microbiological pollution (2 strip)* is the territory surrounding the strip of strict protection. The size of the strip, depending on average annual abstraction with 15 years prediction, can be defined or calculated (computed). If average annual abstraction varies from 10 to 100 m^3/day (with the exception to natural mineral water sources) the size of the strip should be 50 m. diameter from abstraction point (geometric criteria). For higher abstraction rates (more 100 m^3/day) the size of the strip should be calculated (computed) using potential pollutant travel time (taking in to account, that pollutant reached aquifer) to abstraction point 200 days for confined aquifers and 400 days for shallow aquifers (time criteria).

Application of protective measures and land use, in order to protect the resource from a qualitative and quantitative point of view, depends on natural protection level (vulnerability) of the primary aquifer.

The *strip for protection against chemical pollution (3 strip)* is usually calculated or computed (taking in to account, that pollutant reached aquifer) using time criteria where potential pollutant travel time are 25 years (10,000 days). Two strips (strip 3a and strip 3b), with different protection measures applied, should be delimited (Fig. 8.1a, b). If recharge of drinking water source is secured by water flow inside aquifer (Fig. 8.1a) and no capture zone forms on land surface (no relation with shallow groundwater, strip 3b), the absence of abandoned boreholes should be controlled and direct discharge of waste water are forbidden.

If shallow groundwater aquifers take part in recharge of primary aquifer (Fig. 8.1b) and capture zone forms on land surface (strip 3a), application of protection measures and land use depends on natural protection level (vulnerability) of the primary aquifer and are less stringent than in strip for protection against microbiological pollution.

For the areas with prospected and approved groundwater resources, which intended to be used in future, strips 2 and 3 should be delineated and land use or other protective measures should be controlled.

In the cases of individual drinking water supply only strip of strict protection (1 strip) is obligatory and it should be not less than 5 m in diameter. The delineation of strips 2 and 3 of protection zone, are not required.

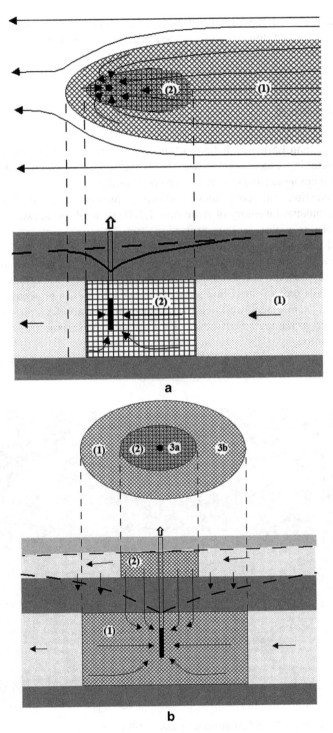

Fig. 8.1 (**a**) Recharge (1) and capture (2) areas of drinking water source in primary aquifers. (**b**) Capture area of drinking water source in primary (1) and in shallow (2) aquifers

According methodology, described above, all aquifers, used for water abstraction, in the territory of Lithuania are grouped according "sensitivity" to pollution: "closed" (I group), "semi-open" (II group) and "open" (III group).

8.2   Assessment

In 2010 more than 1,500 well fields (and single wells) were used to provide population with drinking water. For 650 well fields the sanitary protection strips were calculated or computed and required reports presented to Geological Survey. Totally sanitary protection zones cover about 1,300 km^2, which means about 2% of the territory of Lithuania. Inventory of more than 1,100 potential groundwater pollution sources, existing within sanitary protection strips, allowed assessing of possible impact to drinking water chemical composition [1, 2].

Assessment of possible or existing impact were carried out using indirect (no measurements) and direct (including measurements) methods. Monitoring data and modelling tools prove – strict limitation of human activities in protection strips of "closed" and "semi-open" well fields are redundant measure. Exceptions are well fields situated in the regions of sandy plains not protected from diffuse and point sources pollution.

Examples of negative impact of uncontrolled urbanization to drinking water quality of "open" well field are presented in Figs. 8.2–8.4.

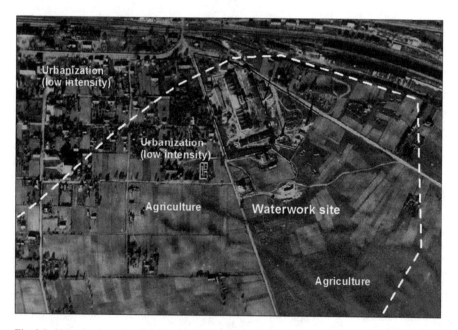

Fig. 8.2 Urbanization of well field surrounding area in 1960–1961

Fig. 8.3 Surrounding area of well field in 2004–2006

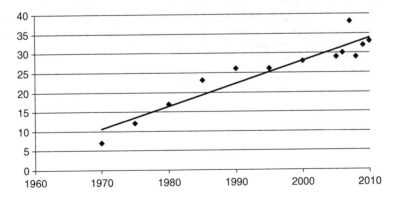

Fig. 8.4 Upward trend of nitrate in well field Lentvaris

The wells for Lentvaris town public water supply were drilled in 1970.

The well field use very vulnerably, water table aquifer consisting of coarse sand and gravel, which lie in 30–40 m depth. In that time the surrounding area were used for low intensity agriculture (gardens, grasslands, etc.) (Fig. 8.2). Only 5–8% of well field capture zone were covered by private houses. The population in that time and today are not connected to public water supply and sewage treatment system. The well field protection zone was not established.

From 1990 the "boom" of private buildings started in the capture zone of the well field (Fig. 8.3).

In 2004–2006 about 35% of the area is covered by private houses without centralized sewage treatment network, however the land use in catchment left in the similar level as in 1960.

The groundwater monitoring in that area started from the beginning of well field operation. In the 1970 concentration of nitrates in drinking water was close to background values of 3–5 mg/l (Fig. 8.4). Gradual urbanization, and indirect input to groundwater pollutants caused upward trend not only of nitrogen compounds, but chlorides and sulphates as well. Today nitrate concentration in drinking water approaching to 40 mg/l, in several wells it exceeds 50 mg/l (Fig. 8.4).

The results of this study are planed to be used for improvement of legislation regulating well head protection areas. Different approaches are planed depending aquifers used for water supply [2]:

- I group (Closed) – no restrictions;
- II group (Semi-open) – case-by case, depending on hydrogeological conditions of each well field;
- III group (Open) – more strict limitation of human activity, investigation (or monitoring data), evaluation, remediation (if risk for drinking water quality).

References

1. Sugalskiene J, Kadunas K (2010) Contaminated sites in well head protection areas: impact assessment to drinking water quality. Technical report. Lithuanian Geological Survey, Lithuanian, p 46
2. Sugalskiene J, Kadunas K (2011) Impact of pollution sources situated in waterworks sanitary protection zones (SPZ) on the quality of drinking water. Lithuanian Geological Survey Annual report 2010, Vilnius, p 44–46

Chapter 9
Advances in GIS-Based Approaches to Groundwater Vulnerability Assessment: Overview and Applications

Alper Elçi

Abstract The concept of groundwater vulnerability assessment is a key component of integrated watershed management. A wide range of approaches for assessing groundwater vulnerability were developed based on identified factors affecting the transport of contaminants in the vadose zone. These approaches can be divided into three major categories: (1) Overlay and index methods which are based on overlaying maps of factors contributing to contamination and subsequently assigning numerical scores or ratings to develop a range of vulnerability classes; (2) process-based methods based on mathematical contaminant transport models; and (3) statistical methods that infer relationships with areas where contamination has already occurred. Many of these approaches are based on a GIS. So-called hybrid methods that involve the combination of approaches are also used and are currently a topic of research. The first objective of this study is to overview recent advances in the GIS-based approaches to groundwater vulnerability assessments. Recent advances in GIS-based vulnerability assessment approaches offer ways to perform more reliable assessments. The integration of actual groundwater quality data in the assessment process, consideration of land use patterns and the use of vigorous optimization schemes are examples of recent improvements in this area. Another objective is to present a summary of a case study demonstrating the application of an optimization procedure for the vulnerability mapping method DRASTIC. The study area is the Tahtalı stream catchment, a major drinking water reservoir supplying 1950 L/s of drinking water to the city of Izmir, the third largest city in Turkey.

A. Elçi (✉)
Department of Environmental Engineering, Dokuz Eylül University,
Tinaztepe Campus, 35160 Buca-Izmir, Turkey
e-mail: alper.elci@deu.edu.tr

F.F. Quercia and D. Vidojevic (eds.), *Clean Soil and Safe Water*, NATO Science for Peace and Security Series C: Environmental Security, DOI 10.1007/978-94-007-2240-8_9,
© Springer Science+Business Media B.V. 2012

9.1   Introduction

With a changing climate and increasing water scarcity, groundwater is a 'hidden resource of water' that can prove very valuable during shortages of surface water resources. On a global scale, according to available data [23] half of the world's drinking water supplies come from groundwater. In arid and semi-arid areas, the dependency on groundwater is even more profound; in some cases it is the only reliable water resource. However, it is also a resource under continuous environmental stress due to many potential contamination sources originating from urbanization, industrial and agricultural activities. Furthermore, it is important to take preventive measures to protect groundwater since the remediation of polluted groundwater is a costly process. Therefore, the protection of groundwater resources from pollution and their sustainable use are essential components of groundwater management.

Although the importance of groundwater has long been acknowledged, the vulnerability to contamination has been recognized only in the recent couple decades. Groundwater was perceived as protected from contamination as a result of human activities by the topsoils and geological medium above the water table. However, it is now known that soils and other layers between the contamination source and the groundwater have a limited capacity to filter and retain contaminants. For this reason, the importance of the concept of groundwater vulnerability has grown, and hence the development of vulnerability assessment methods. In the context of integrated management of contaminated land, groundwater vulnerability assessment is a process that aims to prevent or minimize contamination of our drinking water sources. In fact, site selection decisions for facilities that are likely to contaminate the groundwater should be based on groundwater vulnerability mapping. Since all aquifers can be regarded as vulnerable, it is important to determine aquifers that are more vulnerable than others. Therefore, the assessment of groundwater vulnerability is an essential component in the management of water resources and is one of the preventive measures to protect them.

Due to many potential pollution sources from industry, agriculture and urbanization, groundwater is a resource prone to contamination, in particular groundwater transmitted in unconfined surficial aquifers. The concept of groundwater vulnerability can have different definitions and meanings in different contexts; in the conventional sense it is a dimensionless and an intrinsic property of the groundwater system that defines the likelihood of breakthrough of any contaminant released at the land surface. It cannot be measured directly and must be derived from information of various hydrogeological and morphological properties. Intrinsic vulnerability is independent of contaminant characteristics. While the location of the source of groundwater contamination may be above or under the land surface, the concept of vulnerability typically relates to anthropogenic contaminants released above the water table at or near the land surface. Common examples for this type of groundwater contamination are agricultural activities, septic tanks and sewer lines in residential areas, leakages from above ground waste ponds, pipelines or storage tanks, infiltration from contaminated surface water and leakages from roads and railways. Naturally occurring groundwater contamination due to for example ore-bearing deposits (geogenic contamination source), saltwater intrusion or peat deposits is out of the context of groundwater vulnerability. The concept of groundwater vulnerability can also be defined as contaminant specific, where in addition to hydrogeological properties of the groundwater system, contaminant properties and the specific mobility of the contaminant of interest are also included in the vulnerability assessment process. Vrba and Zaporozec [22] define specific vulnerability as the risk of pollution due to potential impact of specific land uses and contaminants.

A wide range of approaches for assessing groundwater vulnerability have been developed based on identified factors affecting the transport of contaminants released at or near the land surface. These approaches can be divided into three major categories: (1) overlay and index methods which are based on overlaying maps of factors contributing to contamination and subsequently assigning numerical scores or ratings to develop a range of vulnerability classes; (2) process-based methods based on mathematical contaminant transport models; and (3) statistical methods that infer relationships with areas where contamination is already present. Process-based methods use models that describe physical, chemical and biological processes to quantify the movement of water and contaminants from the land surface through the vadose zone. These methods vary in complexity from simple 1-D unsaturated zone transport models to coupled, unsaturated–saturated, multiphase, two- or three-dimensional models. Also, they require an extensive coverage of data. Statistical methods require databases of contamination which are statistically linked to factors contributing to groundwater vulnerability using methods such as multivariate regression, discriminant analysis and factor analysis. So-called hybrid methods that involve the combination of the above mentioned approaches are recently favored and are currently a topic of research. Many of the approaches in these categories are executed within a GIS framework, especially the overlay and index methods are performed exclusively using GIS applications.

This chapter provides an overview of the assessment of groundwater vulnerability using GIS-based overlay and index methods. This introduction is followed by a description of the methodology and data requirements for these methods. After a discussion of benefits and limitations of GIS-based assessment methods, recent advances in vulnerability assessment methods are presented. Furthermore, an overview of a case study demonstrating the application of an example optimization procedure for the vulnerability mapping method DRASTIC is presented.

9.2   GIS-Based Overlay and Index Methods: Methodology and Data Requirements

A variety of techniques were used in the past to estimate ground water vulnerability to contaminants. The nature of techniques for vulnerability estimation varies from simple but data-intensive overlay and index approaches to hybrid methods with integrated complex simulation models. The selection of the best method of vulnerability estimation is dependent on data availability, the nature of the contaminant, types of contamination sources, and the geochemical characteristics of the contaminant. Overlay and index methods involve assigning numerical ratings to hydrogeological and physiological attributes and subsequent weighted combination (overlaying) of various rating maps of these attributes to develop a range of final vulnerability classes. The proliferation of GIS technology made it increasingly easy to apply overlay and index methods.

The most widespread overlay and index methods are based on the DRASTIC method [1], which is named after the seven parameters considered in the method: (1) **D**epth to water table; (2) net **R**echarge; (3) **A**quifer media; (4) **S**oil media; (5) **T**opography; (6) **I**mpact of the vadose zone; and (7) hydraulic **C**onductivity of the saturated zone. This method is conceptually accepted by the hydrogeology community and since its inception several modified versions were developed to improve it.

Other methods that are applied to map groundwater vulnerability using GIS are SINTACS [5], GOD [10], AVI [20], EPIK [7], PI [12] and COP [21]. These methods are conceptually very similar to the DRASTIC method and rely on some sort of weighted overlaying of miscellaneous data layers. Although the concept of the method is relatively simple, it is data-intensive and many processing steps may be necessary to prepare primary raw data for subsequent analysis using GIS tools. More on the advantages and drawbacks of the method is discussed later in the chapter.

9.2.1 Preparation of the Parameter Maps

The seven parameters in the DRASTIC method represent the main factors that contribute to the intrinsic vulnerability, or pollution potential of any aquifer. Descriptions of these factors and their relevance to groundwater contamination are presented in the original documentation of the method by Aller et al. [1], and therefore will not be discussed here. A schematic showing the data requirements and the steps of the method is given in Fig. 9.1. For each of the seven parameters, a grid-based map in raster format is prepared which contains the spatial distribution of vulnerability rating values from 1 (least effective on vulnerability) to 10 (most effective on vulnerability). The assignment of rating values is a subjective process that is controlled by the expert performing the vulnerability assessment. This is considered to be a major drawback of this method. The preparation of the parameter rating maps requires the analysis, synthesis and mapping of more or less raw data in different formats, sources and scales. This is accomplished within a GIS framework that enables the user to maintain databases, process, analyze and map the data. Furthermore, professional judgment and expertise in contaminant hydrology and GIS operations are essential in the preparation of these maps.

Raw data used in groundwater vulnerability assessments come from various sources and in different formats and scales. As illustrated in Fig. 9.1, the most common data

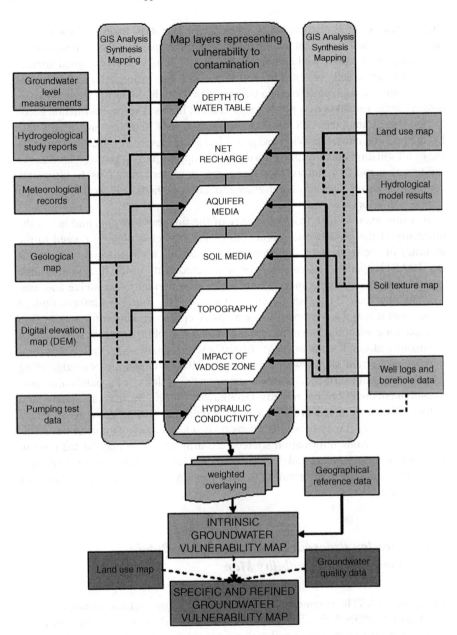

Fig. 9.1 Data requirements and process steps for the GIS-based groundwater vulnerability assessment based on the DRASTIC method. *Dashed connectors* represent optional processes in cases of data scarcity

are groundwater level measurements, meteorological reports, hydrogeological study reports, geological maps, soil texture maps, land use maps, digital elevation models, well pumping test data, well and borehole logs and results of hydrological simulations. These are usually obtained from reports of previous hydrogeological studies, or existing data are augmented by data collection in the field. Some data may be available in tabular format or may not be mapped at all. In this case the information needs to be entered into the GIS database for subsequent mapping and processing. Digitizing, projection, rectifying and geo-referencing may also be necessary to make use of information obtained in the form of hardcopy maps, e.g. geology, land use or soil texture maps. Also, information that comes in vector format must be ultimately converted to raster format as the overlay and index methods only work when raster format map layers are used.

It is important to note that the quality of the final product map and hence the reliability of the groundwater vulnerability assessment is directly related to the accuracy of the input data. Any significant uncertainties associated with any of the DRASTIC parameters can deteriorate the credibility of the final product, the groundwater vulnerability map. Therefore, it is important to scrutinize and analyze the input data, if necessary verify using secondary sources of information. In some cases it may be warranted to fix maps or update existing information. One example for correcting data in the groundwater vulnerability process is to identify erroneous sinks and peaks in DEMs (digital elevation models). The DEM is used to calculate terrain slopes which affect groundwater contamination originating near or at the land surface. Another example is the updating of groundwater level measurements with current records. Groundwater levels fluctuate seasonally and in time. Therefore dated groundwater level data may not reflect current conditions. Sometimes it may occur that maps related to the same parameter but originating from various information sources are different. This may be the case in particular with soil texture and geological maps. Here it is necessary to synthesize and combine the given information to derive the most accurate representation of the parameter.

9.2.2 Overlay Procedure and Indexing to Obtain the Final Vulnerability Map

The seven DRASTIC parameters are assigned weighting coefficients ranging from 1 to 5, which reflect their relative importance to the vulnerability of the aquifer. In the first iteration of the process, these weights are determined subjectively by the expert who performs the vulnerability assessment. However, these weights can be subsequently refined based on statistical correlation of actual groundwater quality data with the obtained vulnerability index values, which leads to the production of a refined, specific groundwater vulnerability map (Fig. 9.1). A case study illustrating this procedure is provided in Sect. 9.4.

Typical weights used in the DRASTIC method are $\lambda_D = 5$; $\lambda_R = 4$; $\lambda_A = 3$; $\lambda_S = 2$; $\lambda_T = 1$; $\lambda_I = 5$ and $\lambda_C = 3$. The vulnerability index is then calculated by linearly summing all seven parameters according to the following equation:

$$\text{DRASTIC index} = D_r\lambda_D + R_r\lambda_R + A_r\lambda_A + S_r\lambda_S + T_r\lambda_T + I_r\lambda_I + C_r\lambda_C \quad (9.1)$$

where $D_r, R_r, A_r, S_r, T_r, I_r$ and C_r are the rating values for the seven parameters affecting groundwater vulnerability and λ the corresponding weight coefficients. The result is a raster map containing vulnerability index values with a possible range of 23–230 which is directly dependent on the assignment of ratings and weight coefficients for the parameters. Higher index values represent areas classified as very susceptible to groundwater contamination, whereas lower index values imply less susceptible areas. The index map is then typically classified into ranges of index values by using arbitrary or statistically meaningful thresholds. It is often reasonable to group the cells into five classes, representing "most vulnerable", "very vulnerable", "moderately vulnerable", "less vulnerable" and "least vulnerable". Other classification options include color-coding the index based on the original DRASTIC method or normalizing the index so that it is rescaled on a 1-to-100 scale, thereby expressing the vulnerability in percentage.

9.2.3 Advantages and Limitations of Gis-Based Overlay and Index Methods

The overlay and index methods are the most commonly and also the earliest methods used in the assessment of groundwater vulnerability. They have some significant advantages; firstly, they have become popular because the methodology is fairly straightforward that can be easily implemented with any GIS application software. The concept of overlaying data layers is easily comprehended even by less experienced users. Also, the data requirement can be considered as moderate since nowadays most data comes in digital format. Hydrogeological information is either available or can be estimated using relevant data. However, the challenge is usually to analyze and synthesize data coming from various sources and ensure its validity and accuracy. The goal in data preparation should be to bring the information to a common ground to enable the use in the vulnerability assessment process. According to a study that examines the DRASTIC method [18], the selection of many parameters and their inter-dependence decrease the probability of ignoring some important parameters, and thereby limit the occurrence of incidental errors. Consequently, these methods give relatively accurate results for extensive areas with a complex geological structure. Lastly, the product of this approach can be easily interpreted by water-resource managers and can be incorporated into decision-making processes. Even a simple visual inspection of the vulnerability map can reveal important contamination hotspots.

Probably the most important and obvious disadvantage of these methods that is raised by scientists and experts is the inherent subjectivity in the determination of the rating scales and the weighting coefficients. These were originally established using the Delphi technique, i.e. an interactive discussion and surveying of a panel of selected independent experts. Although the worldwide use of the same typical ratings and weights ensures the production of comparable vulnerability maps on a regional scale, the selection of the numerical values and the scale of ratings and weights remain disputable.

Criticism is also expressed for the selection of the parameters that affect groundwater vulnerability. It is claimed by some experts that redundant parameters are factored into the final vulnerability index. It is in fact evident that some parameters are not independent but somewhat correlated. For example the depth to groundwater is also a function of several other vulnerability parameters such as net groundwater recharge, topography and hydraulic conductivity of the aquifer. Also, one can argue that groundwater recharge is inversely proportional to the terrain slope (topography).

Another disadvantage of GIS-based methods is the indirect consideration of favorable factors in the aquifer such as attenuation or degradation, dilution and contaminant travel time from the pollution source to the water table. The integration of these processes to the GIS-based methods would require the use of sophisticated and data-intensive groundwater flow and contaminant transport models which would probably result in vulnerability maps that are scientifically more defensible, however at a significantly higher cost.

Furthermore, different overlay and index methods applied to the same system can yield significantly dissimilar results (e.g. [11]). As the groundwater vulnerability index is relative in nature rather than absolute, comparison of maps obtained with different methods can be inconclusive.

9.3   State-of-the-Art in Groundwater Vulnerability Mapping

In recent years, many modifications and improvements of the DRASTIC approach were published. The common objective of these modifications is to address several drawbacks of the overlay and index methodology. As it is mentioned earlier in this chapter, groundwater vulnerability assessment methods are traditionally grouped into overlay and index methods, process-based methods and statistical methods. Each group of methods has their strengths and weaknesses. However, by combining components of these approaches, improved and more defensible methods can be developed. These methods are referred to as hybrid methods [2, 9].

Identified and criticized issues with the overlay & index methodology was mentioned in the previous section. Several approaches are developed to mitigate shortcomings of these methods, one of them being the development of hybrid methods. One of these issues is the arbitrariness of vulnerability factor selection and the correlations among factors. This issue of subjectivity can be partly mitigated by selecting the vulnerability factors based on quantitative judgment. For example, analyzing the sensitivity of vulnerability index values to the systematic exclusion of each parameter in question can eliminate redundant factors that can be subsequently omitted in the assessment process. However, one can rightfully argue that such a sensitivity analysis would make the entire assessment procedure more complicated and time consuming. A study by Babiker et al. [3] demonstrates the use of sensitivity analysis in the selection of factors. Moreover, the inclusion of additional factors such as land use/land cover (LULC) information, spatial patterns of irrigation and likely contamination sources is another way to use more relevant factors and to increase the reliability of the vulnerability assessment (e.g. [14]) The issue of subjectivity in the determination of vulnerability factor weights is mitigated by optimizing the overlay and index procedure. This can be accomplished by introducing actual groundwater quality data into the entire assessment process such as for example nitrate concentration measurements. Nitrate is a diffuse source pollutant generally originating from agricultural activities and domestic wastewater. It is transported in groundwater without any retardation, i.e. it flows with the same velocity as water. The correlation of the occurrences of contamination with the corresponding vulnerability index values are determined to validate the vulnerability assessment. Based on this evaluation necessary adjustments of the ratings and weight coefficients to increase the correlation are made. Panagopoulos et al. [17] present a good example of an optimization approach for the DRASTIC method.

Physical, chemical and, if relevant, biological processes occurring in the subsurface can be considered to enhance the vulnerability assessment. Neukum and Azzam [16] present an approach using numerical simulation of water flow and solute transport with transient boundary conditions and new vulnerability indicators. They used a 1-D finite element model to determine breakthrough of contaminants from the land surface to the water table. A similar study by Connell and Van den Daele [6] illustrates an approach derived from a series of analytical and semi-analytical solutions to the advection–dispersion equation that includes root zone and unsaturated water movement effects on the transport process. The steady-state form of these equations provides an efficient means of calculating the maximum concentration at the water table and therefore has potential for use in vulnerability mapping. Furthermore, consideration of well capture zones combined with an overlay and index method can further advance the existing approaches (e.g. [15]). This kind of an approach represents also a tool that can aid in the evaluation wellhead protection areas based not only on the conventional time-of-travel concept but also on potential contaminant sources and land use practices within the capture zone.

9.4 Case Study: Tahtalı Stream Basin in Izmir-Turkey

A summary of a case study is presented in this section with the aim of demonstrating the practical application of an advanced GIS-based groundwater vulnerability assessment method. The original version of DRASTIC is implemented and ArcGISv9.3 is used as the GIS software. The selected study area is a catchment for a major drinking water reservoir for the city of Izmir, the third largest city in Turkey. The objectives of this study are (1) to obtain a groundwater vulnerability map for the Tahtalı stream basin; (2) optimize this vulnerability map; and (3) evaluate reliability of the optimized vulnerability map with actual contamination data.

9.4.1   Description of the Study Area

The Tahtalı dam reservoir (38°08′ N; 27°06′ E) is located 40 km south of Izmir and meets about 36% of the city's total water demand (Fig. 9.2). The catchment has an area of 550 km² and it is a sub-basin of the larger K. Menderes river basin. The climate of the region is typical Mediterranean and the long-term mean annual precipitation over the area is recorded as 690 mm. Most of the precipitation occurs in the period between October and May, whereas the rest of the year is relatively dry. A total of 44 ephemeral creeks and their tributaries drain through the catchment area and feed the Tahtalı reservoir. The major inflows to the reservoir are from the north via Tahtalı and Şaşal streams. The Şaşal stream contributes 25% whereas the Tahtalı stream contributes 75% to the total inflow. The discharges of the other four inflowing streams are negligible [4]. According to the 2008 census data [19], 68,000 inhabitants live within the boundaries of the Tahtalı catchment. Based on analysis of satellite imagery, the land use distribution of the area is 42.1% forests, 31.8% agricultural areas, 3.1% water bodies, 1.8% residential and 0.2% industrial areas. Quaternary aged alluvial deposits, Neogene age flysch, clayey limestone, allochthonous limestone, conglomerate and tuff formations comprise the geological structure of the study area.

   The catchment is under several natural and anthropogenic environmental stresses including land use alteration, groundwater over-exploitation and droughts despite actively implemented administrative protection measures against pollution. The groundwater in the catchment is prone to contamination originating from domestic and some industrial pollution sources located mostly in the medium- and long-distance buffer zones of the catchment. On the other hand, there are many wells drilled in the surficial aquifer, which provide the much needed irrigation water for agriculture. Furthermore, excessive withdrawal from the surficial aquifer due to increase in population and also periodic droughts that have become more pronounced and sustained due to climate change pose a serious threat to the quantity and quality of groundwater resources.

9.4.2   Data Processing and Preparation of DRASTIC Rating Maps

Information about the hydrogeology of the study area is obtained from various sources in the form of field data, reports, digital and hardcopy maps.

   The seven DRASTIC parameter map layers are prepared with a raster resolution of 100×100 m. UTM coordinates (zone 35S) and the European 1950 datum is chosen for all map layers. The following raw data is used in this study:

* Seasonal groundwater level measurements at 51 monitoring points
* 76-m resolution DEM from NASA's SRTM database
* Geology map
* Soil texture map
* Land use map

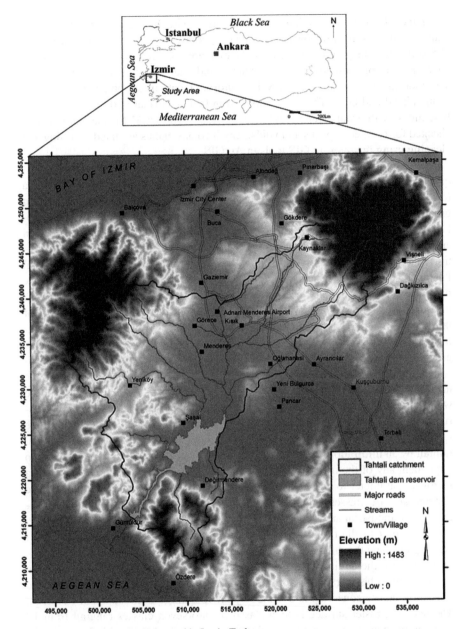

Fig. 9.2 Tahtalı stream basin located in Izmir, Turkey

- Well logs and well specific capacity data for over 100 drilling locations
- Miscellaneous reports about previous hydrogeological studies in the area
- Cartographic data such as highways, locations of residential areas and stream network

All the information is available in digital format with the exception of the geology map. This map is in hardcopy and is scanned, georeferenced and digitized using the GIS software. The geological areas on the map are digitized as polygons which are later checked for geometrical inconsistencies and finally converted to raster format. The coordinates of groundwater level monitoring locations are obtained in the field using a handheld GPS. These are imported in tabular format into the project database along with groundwater level measurements. The DEM for the study area is checked for erroneous peaks and voids. Any inconsistencies are fixed by filling and cleaning using the relevant GIS tools in ArcGIS.

The rating maps for each parameter are produced by reclassifying information of quantitative data (e.g. groundwater depth, hydraulic conductivity) or qualitative data (e.g. aquifer and soil type) based on a scale from 1 to 10, depending on the degree of groundwater vulnerability. Groundwater depth is calculated using cokriging interpolation on seasonal groundwater levels at 31 monitoring points [8]. The collocated auxiliary data used in the cokriging process is selected as the ground surface elevation data from the DEM of the study area. It is important to note that due to seasonal fluctuation of the water table the least favorable situation with respect to vulnerability must be taken into account. Since the water table is highest in the rainy winter months, measurements representative of the winter conditions are used in the preparation of the groundwater depth parameter map. Rating values are assigned to each class ranging from 1 (deep groundwater, thus minimum impact on vulnerability) to 10 (shallow groundwater, thus maximum impact on vulnerability).

Net recharge used in the context of groundwater vulnerability assessment is the total amount of precipitation water that infiltrates from the ground surface to the water table. This parameter can be estimated using hydrological precipitation-runoff models, field experiments or simply by multiplying the difference of the spatial distribution of evapotranspiration and the spatial distribution for mean annual rainfall by an infiltration coefficient. However, the necessary data is not available to implement any of the mentioned methods. Instead the land use map was used as a surrogate parameter to derive the recharge map. Rating values are assigned to five distinct land use classes, ranging from 10 (agricultural areas) to 1 (highways, residential and industrial areas and water bodies).

The aquifer media ratings map is primarily obtained from rigorous analysis of well log data and the digitized geology map. The latter is used in particular to delineate zones for the aquifer media ratings map. Rating values are assigned to seven aquifer media classes ranging from 10 (karstic limestone) to 2 (tuff). Similarly, the impact of vadose zone map is obtained by interpolating point data from well logs. The resulting raster surface is then classified into seven classes ranging from 10 (maximum impact on vulnerability due to absence of impermeable layers in the vadose zone) to 3 (least impact to vulnerability due to fractured formations or relatively permeable layers).

The soil type parameter map is obtained directly from an available digital version of a soil texture map for the Tahtalı stream basin. The different soil types are assigned ratings based on the potential ability of the soil to filter out contaminants by sorption and biochemical removal processes. These processes are usually governed by the clay and organic matter contents of the soil, respectively. Assigned ratings ranged from 10 (bare rock, gravel) to 1 (clay).

The impact of topography on vulnerability is related to the terrain slope. Slope is calculated in terms of percentage using the resampled version of the original digital elevation data. Resampling is necessary to be consistent with the selected 100×100 m resolution since the DEM had a different resolution. The calculated slope is classified into seven classes based on the prescribed rating scheme of the original DRASTIC method.

Hydraulic conductivities for the saturated aquifer are estimated using specific capacity data. This data is usually obtained after completion of a well drilling and is provided in the well log. The inverse distance weighted interpolation scheme is selected to generate the spatial distribution of hydraulic conductivity from point data. The resulting raster map is further classified into four classes based on typical rating values ranging from 1 to 4.

In the conventional DRASTIC method, the seven vulnerability parameters are assigned weighting coefficients ranging from 1 to 5, which reflect their relative importance to the vulnerability of the aquifer. These weights are determined subjectively by the expert who performs the vulnerability assessment. However, these weights can be subsequently refined based on statistical correlation of actual groundwater quality data with the obtained vulnerability index values, which leads to the production of an optimized groundwater vulnerability map, as it was done in this study. The "raw" DRASTIC vulnerability index map is computed according to the linear addition of the parameters given in Eq. 9.1. Weight coefficients of the original DRASTIC method was used: $\lambda_D = 5$; $\lambda_R = 4$; $\lambda_A = 3$; $\lambda_S = 2$; $\lambda_T = 1$; $\lambda_I = 5$; $\lambda_C = 3$.

9.4.3 Optimization of Groundwater Vulnerability Maps with Nitrate Data

The raw, unoptimized groundwater vulnerability map for the study site is obtained by weighted overlaying of the seven parameter rating maps using original DRASTIC weighting coefficients. Following this step, the map is refined by optimizing the parameter weighting coefficients using groundwater quality data. Nitrate is selected as quality data which is considered to be representative of land surface (or near surface in the case of septic tank leakages) originating, conservative contaminant. Nitrate concentrations that were measured at 31 sampling points in a previous study [13] are used for this purpose. Nitrate concentrations and sampling points are depicted in Fig. 9.3.

The optimization of parameter weighting factors is achieved first by calculating the Spearman's rho correlation coefficients of each vulnerability parameter with the nitrate concentration at the 31 sampling points. Consequently, seven correlation coefficients are determined. These coefficients are rescaled to a scale of 1–5 to obtain the optimized parameter weighting coefficients for each vulnerability parameter. In the case when any correlation coefficient is statistically insignificant, the related

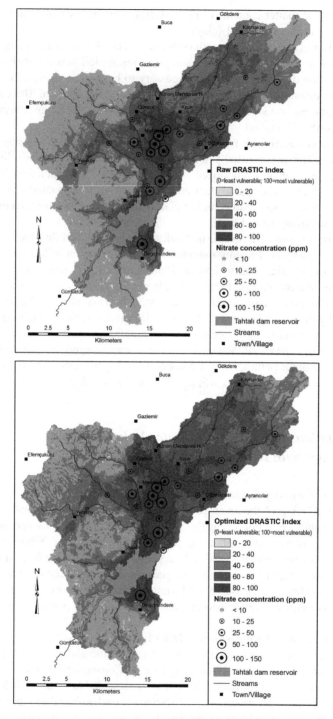

Fig. 9.3 Raw groundwater vulnerability map of the Tahtalı stream basin (*top*) and optimized groundwater vulnerability map using optimized DRASTIC weighting coefficients (*bottom*)

Table 9.1 Original and optimized weighting coefficients and correlations of vulnerability parameters with nitrate data

Vulnerability parameter	Original weighting coefficient	Spearman's rho correlation coefficient ($p < 0.05$)	Optimized weighting coefficient
Groundwater depth (D)	5	0.679	2.58
Net recharge (R)	4	0.607	1.10
Aquifer type (A)	3	0.732	3.67
Soil type (S)	2	0.750	4.03
Topography (T)	1	0.760	4.24
Impact of vadose zone (I)	5	0.797	5.00
Hydraulic conductivity (C)	3	0.602	1.00

parameter is simply ignored and taken out of Eq. 9.1. Using the adjusted parameter weighting coefficients, an optimized version of the groundwater vulnerability map is created. As part of the performance evaluation of the process, the correlations of raw and optimized vulnerability index values with actual nitrate concentrations are used as criteria to verify the accuracy of the obtained vulnerability maps. Thereby, it is also possible to quantify the impact of the optimization process on the overall outcome.

9.4.4   Results

Correlation coefficients of vulnerability parameters with nitrate concentrations and the modified weighting factors are summarized in Table 9.1. Based on these results, all parameters had statistically meaningful correlations with nitrate data. The vadose zone impact parameter and the saturated hydraulic conductivity have the strongest and weakest correlation, respectively. The weightings of groundwater depth, net recharge and hydraulic conductivity are decreased in the overall vulnerability assessment process. The remaining parameters are assigned higher weightings and hence become more pronounced vulnerability factors. The unoptimized, raw and the optimized groundwater vulnerability map are provided in Fig. 9.3.

Groundwater vulnerability indices for both maps are normalized based on Eq. 9.2 to attain comparability.

$$X_n = \frac{X - X_{min}}{X_{maks} - X_{min}} \times 100 \qquad (9.2)$$

X_{min} and X_{max} represent the lowest and highest index values calculated for the map, respectively, X the raster index value and X_n the normalized vulnerability index. Also, to facilitate the interpretation of the maps, index values are reclassified into five classes; the light tones in the map depict the areas that are less susceptible to groundwater contamination and the dark tones the areas where contamination should be expected, if a contamination source is present.

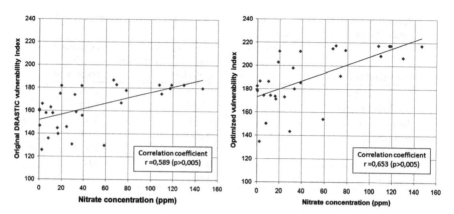

Fig. 9.4 Correlation between vulnerability index and nitrate concentration; (*left*) original DRASTIC vulnerability index; (*right*) optimized vulnerability index

Pearson correlation coefficients between measured nitrate concentrations and corresponding vulnerability ratings increase from 0.589 to 0.653 by optimizing the layer weights (Fig. 9.4). It can be concluded that the optimization improves the correlation by 11%. Furthermore, it can be calculated using the optimized map that on an aerial basis 44% of the watershed is vulnerable to groundwater contamination, whereas 30% and 26% is moderately and less vulnerable, respectively. In particular the central part of the catchment is dominated by high vulnerability index values. This is in part due to the fact that the water table is shallow in the alluvial plains around the town of Menderes. Also, the geology in the central part of the catchment area is dominated by alluvial deposits, another negative factor with respect to vulnerability. Higher elevations of the study area, e.g. the northeast, south and west, are generally classified as less vulnerable due in part by deep groundwater and high terrain slopes.

9.5 Conclusions

Many different methods are applied to create groundwater vulnerability maps, which are used by water resource authorities in the decision-making process of groundwater wellhead protection zoning or as a preventive measure in contaminated land management. GIS-based groundwater vulnerability assessment methods are viable and useful tools for decision makers who want to take measures to protect groundwater resources from contamination. Most GIS-based methods are widely accepted and used in practice because they are conceptually simple and their application is fairly straightforward. Although the GIS operations during the execution of the assessment are not complicated, it should be noted that meticulous preparation of the parameter rating maps is important. The reliability of the vulnerability maps depends on data and parameter map accuracy and quality. It is necessary to analyze the quality of raw data and maps originating from various sources and then synthesize the information to produce the parameter maps that form the basis of the vulnerability assessment.

Another important fact of the matter is that even with visual inspection of the final vulnerability map certain patterns of groundwater vulnerability can be revealed that can be easily justified with data and observations in the field. However, it should be also noted that these methods give the end user a vulnerability map that is relative in nature rather than absolute. Therefore, if possible, enhancement or optimization of the assessment process should be sought. It was shown in the presented case study that subjectivity issues of the overlay and index assessment approach can be reduced by a simple optimization process which is tied to actual groundwater quality data.

References

1. Aller L, Bennett T, Lehr JH, Petty RJ, Hackett G (1987) DRASTIC: a standardized system for evaluating ground water pollution potential using hydrogeologic settings. Environmental Research Laboratory, Office of Research and Development, U.S. Environmental Protection Agency, Ada
2. Antonakos AK, Lambrakis NJ (2007) Development and testing of three hybrid methods for the assessment of aquifer vulnerability to nitrates, based on the drastic model, an example from NE Korinthia, Greece. J Hydrol 333(2–4):288–304
3. Babiker IS, Mohamed MAA, Hiyama T, Kato K (2005) A GIS-based DRASTIC model for assessing aquifer vulnerability in Kakamigahara Heights, Gifu Prefecture, central Japan. Sci Total Environ 345(1–3):127–140
4. Çalışkan A, Elçi Ş (2009) Effects of selective withdrawal on hydrodynamics of a stratified reservoir. Water Resour Manag 23(7):1257–1273
5. Civita M (1994) Le Carte délia Vulnerabilitâ degli Acquiferi All'inquinamento (Aquifer vulnerability maps to pollution). Pitagora Ed, Bologna
6. Connell LD, Van den Daele G (2003) A quantitative approach to aquifer vulnerability mapping. J Hydrol 276(1–4):71–88
7. Doerfliger N, Jeannin PY, Zwahlen F (1999) Water vulnerability assessment in karst environments: a new method of defining protection areas using a multi-attribute approach and GIS tools (EPIK method). Environ Geol 39(2):165–176
8. Elçi A, Karadaş D, Fıstıkoğlu O (2010) The combined use of MODFLOW and precipitation-runoff modeling to simulate groundwater flow in a diffuse-pollution prone watershed. Water Sci Technol 62(1):180–188
9. Focazio MJ, Reilly TE, Rupert MG, Helsel DR (2002) Assessing ground-water vulnerability to contamination: providing scientifically defensible information for decision makers. U.S. Geological Survey Circular 1224. U.S. Geological Survey (USGS), Reston
10. Foster SSD (1987) Fundamental concepts in aquifer vulnerability, pollution risk and protection strategy. In: Duijvenbooden W, Waegeningh HGV (eds) Vulnerability of soil and groundwater to pollutants. TNO Committee on Hydrogeological Research, The Hague, pp 69–86
11. Gogu R, Hallet V, Dassargues A (2003) Comparison of aquifer vulnerability assessment techniques. Application to the Néblon river basin (Belgium). Environ Geol 44(8):881–892
12. Goldscheider N (2005) Karst groundwater vulnerability mapping: application of a new method in the Swabian Alb, Germany. Hydrogeol J 13(4):555–564
13. Ileri B, Gündüz O, Elçi A, Şimşek C, Alpaslan MN (2007) GIS integrated assessment of groundwater quality in Tahtali Basin. In: Proceedings of 7th National Environmental Engineering Congress, Izmir (in Turkish)
14. McLay CDA, Dragten R, Sparling G, Selvarajah N (2001) Predicting groundwater nitrate concentrations in a region of mixed agricultural land use: a comparison of three approaches. Environ Pollut 115(2):191–204

15. Nobre RCM, Rotunno Filho OC, Mansur WJ, Nobre MMM, Cosenza CAN (2007) Groundwater vulnerability and risk mapping using GIS, modeling and a fuzzy logic tool. J Contam Hydrol 94(3–4):277–292

16. Neukum C, Azzam R (2009) Quantitative assessment of intrinsic groundwater vulnerability to contamination using numerical simulations. Sci Total Environ 408(2):245–254

17. Panagopoulos GP, Antonakos AK, Lambrakis NJ (2006) Optimization of the DRASTIC method for groundwater vulnerability assessment via the use of simple statistical methods and GIS. Hydrogeol J 14(6):894–911

18. Rosen L (1994) A study of the DRASTIC methodology with emphasis on Swedish conditions. Ground Water 32(2):278–285

19. TURKSTAT (2008) Databank for population, housing and demography. Turkish Statistics Institution. http://www.tuik.gov.tr

20. Van Stempvoort D, Ewert L, Wassenaar L (1993) Aquifer Vulnerability Index (AVI): a GIS compatible method for groundwater vulnerability mapping. Can Water Resour J 18:25–37

21. Vías J, Andreo B, Perles M, Carrasco F, Vadillo I, Jiménez P (2006) Proposed method for groundwater vulnerability mapping in carbonate (karstic) aquifers: the COP method. Hydrogeol J 14(6):912–925

22. Vrba J, Zaporozec A (1994) Guidebook on mapping groundwater vulnerability. IAH International Contributions to Hydrogeology, vol 16. Leiden, The Netherlands

23. Zebidi H (1998) Water, a looming crisis? In: Proceedings of the international conference on world water resources at the beginning of the 21st century (UNESCO, Paris, 3–6 June 1998). UNESCO International Hydrological Programme – Technical Documents in Hydrology, vol 18. UNESCO, Paris

Chapter 10
European Ground Water Geochemistry Using Bottled Water as a Sampling Medium

Alecos Demetriades, Clemens Reimann, Manfred Birke, and The Eurogeosurveys Geochemistry EGG Team

Abstract To obtain a first impression of the geochemistry and quality of European ground water bottled mineral water was used as a sampling medium. In total, 1,785 bottled waters were purchased from supermarkets of 40 European countries, representing 1,247 wells/drill holes/springs at 884 locations. All bottled waters were analysed for 72 parameters at the laboratories of the Federal Institute for Geosciences and Natural Resources (BGR) in Germany. The geochemical maps give a first impression of the natural variation in ground water at the European scale. Geology is one of the key factors influencing the observed element concentrations for a significant number of elements. Examples include high values of (i) Cr clearly related to the occurrence of ophiolites; (ii) Li (Be, Cs) associated with areas underlain by Hercynian granites; (iii) F (K, Si) related to the occurrence of alkaline rocks, especially near the volcanic centres in Italy, and (iv) V indicating the presence of active volcanism and basaltic rocks. For some elements, the reported concentrations are influenced by bottle material. In general, glass bottles leach more elements (Ce, Pb, Al, Zr, Ti, Hf, Th, and La) to stored water than PET bottles. However, all values observed during the leaching tests were well below the respective maximum admissible concentrations, as defined for drinking water by European Union legislation.

A. Demetriades (✉)
Institute of Geology and Mineral Exploration, Hellas, Athens, Greece
e-mail: ademetriades@igme.gr

C. Reimann
Geological Survey of Norway, 7491 Trondheim, Norway
e-mail: clemens.reimann@ngu.no

M. Birke
Federal Institute for Geosciences and Natural Resources, BGR, Hannover, Germany
e-mail: manfred.birke@bgr.de

The Eurogeosurveys Geochemistry EGG Team
EuroGeoSurveys, Brussels, Belgium

F.F. Quercia and D. Vidojevic (eds.), *Clean Soil and Safe Water*, NATO Science for Peace and Security Series C: Environmental Security, DOI 10.1007/978-94-007-2240-8_10, © Springer Science+Business Media B.V. 2012

10.1   Introduction

The EuroGeoSurveys Geochemistry Expert Group is dedicated to provide high quality databases on the geochemistry of earth materials with which humans are in direct contact with to decision makers, geoscientists, researchers and the public alike.

The Geochemical Atlas of Europe provided the first harmonised pan-European multi-determinand databases on residual soil (top- and sub-soil), humus, stream and floodplain sediments, and stream water [10, 33].

Ground water, although very important was, however, missing from this database. The main reason is that to collect systematically representative ground water samples at the European scale is not an easy task, and may be prohibitively expensive if performed at a high sample density.

It was against this background that the EuroGeoSurveys Geochemistry Expert Group put forward a novel idea that "ground water" samples can be readily bought from supermarkets throughout Europe as bottled mineral water. Therefore, it should be possible to use it as a first proxy for ground water geochemistry and quality at the European scale.

Though the idea was met with some scepticism to begin with, it was finally decided that it was worth a test, because it provided a cost-effective approach.

The results of this project are presented in a geochemical atlas [25], and in a special issue of the Journal of Geochemical Exploration [2], as well as in other publications [11, 21, 28, 29]. In this chapter a concise summary of the project is given with some key results.

10.2   Sampling

Instructions were sent to all members of the EuroGeoSurveys Geochemistry Expert Group, as well as to friends and colleagues travelling to European countries, to purchase from supermarkets as many different bottled mineral water brands as possible.

In case the same bottled water was available with and without gas, both varieties were purchased. If bottled water was marketed in different bottle types (e.g., glass and PET), or in bottles of different colour all varieties were bought whenever possible. To keep shipping costs down 0.5 L bottles were preferentially bought. However, if a brand was exclusively marketed in larger capacity containers of 1, 1.5 or even 5 L these were purchased.

All national bottled water samples were shipped to the laboratory of the Federal Institute for Geosciences and Natural Resources (BGR) in Berlin, Germany, where they were kept refrigerated until analysed. The sampling period started in November 2007 and ended in April 2008. The total number of bottles purchased, and subsequently analysed, was 1,785.

Before shipping, the analytical results recorded on bottle labels were transferred to an Excel worksheet together with other pertinent information.

10.3   Laboratory Analysis and Quality Control

10.3.1   Laboratory Analysis

The bottled water samples were analysed at the chemical laboratory of the Federal Institute for Geosciences and Natural Resources (BGR) in Berlin. Details of sample preparation and the extensive analytical programme are reported by Reimann and Birke [25] and Birke et al. [3]. Thus, only a summary of the methods employed is provided here:

- Inductively coupled plasma atomic quadruple mass spectrometry (ICP-QMS): Ag, Al, As, B, Ba, Be, Bi, Cd, Ca, Ce, Co, Cr, Cs, Cu, Er, Eu, Fe, Ga, Gd, Ge, Hf, Hg, Ho, I, K, La, Li, Lu, Mg, Mn, Mo, Na, Nb, Nd, Ni, Pb, Pr, Rb, Sb, Sc, Se, Sm, Sn, Sr, Ta, Tb, Te, Th, Ti, Tl, Tm, U, V, W, Y, Yb, Zn, Zr;
- Inductively coupled plasma atomic emission spectrometry (ICP-AES): Ba, Ca, K, Mg, Mn, Na, Sr, P, and Si;
- Ion Chromatography (IC): Br^-, Cl^-, F^-, NO_2^-, NO_3^-, SO_4^{2-};
- Atomic fluorescence spectroscopy (AFS): Hg;
- Titration: total alkalinity - HCO_3^-;
- Photometric: NH_4^+;
- Potentiometric: pH, and
- Conductometric: Electrical Conductivity (EC).

10.3.2   Quality Control

A very strict quality control programme was installed and reported by Reimann and Birke [25] and Birke et al. [3]. Below a concise outline is given:

1. Analysis of international reference samples to document the trueness of analytical results, i.e., the river water reference material SLRS-4 from the National Research Council Canada, and the low level fortified standards for trace elements TM-26.3, TM-27.2, TM-28.2 and TM-28.3 from the National Water Research Institute of Canada;
2. Frequent analysis of an in-house project standard (MinWas) to check the accuracy of determined parameters;
3. Frequent analysis of blank samples to detect any contamination issues and to derive reliable detection limits;
4. Frequent analysis of sample duplicates to determine precision of measurements;
5. Comparison of analytical results of this study with those recorded on bottle labels;
6. Determination of a few parameters by two methods (Ba, Ca, K, Mg, Mn, Na, Sr) by ICP-QMS and ICP-AES, and Hg by ICP-QMS and AFS, and
7. Buying a new bottle and re-analysing the bottled water with unusually high results for important parameters whenever possible.

A general problem of analysing so many elements, as in this study, is that there are no suitable reference materials to cover all elements.

The Canadian standards (SLRS-4, TM-26.3, TM-27.2, TM-28.2 and TM-28.3) have the advantage of covering different concentration ranges for a number of elements, and could be used to identify elements that were problematic at low concentrations (e.g., Hf, Nb, Sn, Ta and W), but delivered reliable results at higher values (i.e., over ten times the detection limit).

Overall, certified values, and the generated project results, are well in agreement for most elements. Since, there exist a number of elements that are not covered by any standards; this drawback was covered by evaluating reliability of results with respect to blank values and coefficient of variation.

10.3.3 Detection Limit

The instrument detection limit (IDL) was estimated at three times the standard deviation of sample blank determinations (Table 10.1).

The reported detection limit (RDL) was calculated at ten times the standard deviation of sample blanks. The conservative RDL was used as the cut-off value for all statistical graphics and tables, as well as for producing the distribution maps.

The duplicate analyses were also used to estimate the practical detection limit [6, 8, 9, 27].

10.4 Influence of Bottle Materials and Carbonatisation

10.4.1 Bottle Leaching

Many studies have demonstrated that water samples can be severely contaminated by the material of storage bottles [18, 22, 31], and often extreme cleansing procedures are suggested for sample bottles [32], and/or the use of very expensive special plastic bottles is suggested for water sampling for ultra-trace element analysis. For bottled water a large variety of different containers are on the market (e.g., glass, hard PET, soft PET, aluminium cans, tetrapacs). Bottles have in addition different colours (clear, light and dark green, blue and brown are the most common colours). It must, thus, be expected that each bottle type will have different properties and leaching characteristics, and will influence the stored water quality in some way. There is considerable documentation that due to the low concentrations of most elements in natural water, water sampling and analysis are required to be performed with great care in order to avoid contamination of water samples during sampling, storage, or analysis [24].

Glass bottles are known to leach Pb and Zr to the stored water [23]. More recently, [19, 20, 34–36, 39] have demonstrated that leaching of elements from bottle

Table 10.1 Instrumental analytical method, detection limits, precision, minimum, maximum and median values of bottled water data set used for mapping (n=884), and statutory values

Parameter	Method	Unit	IDL*	RDL**	PDL***	Precision[a]%	Min	Median	Max
Ag	ICP-QMS	µg/L	0.001	0.002	0.002	13	<0.002	<0.002	112
Al	ICP-QMS	µg/L	0.2	0.5	0.2	5	<0.5	1.2	966
As	ICP-QMS	µg/L	0.01	0.03	0.001	10	<0.03	0.24	90
B	ICP-QMS	µg/L	0.1	2	0.2	4	<2	39	120,000
Ba	ICP-QMS	µg/L	0.005	0.1	1[c]	5	0.05	29	26,800
Be	ICP-QMS	µg/L	0.001	0.01	0.005	5	<0.01	<0.01	64
Bi	ICP-QMS	µg/L	0.0005	0.005	0.003	–[d]	<0.005	<0.005	0.69
Br⁻	IC	µg/L	3	3	–	–	<3	35	21,700
Ca	ICP-OES	mg/L	0.005	0.01	–	–	0.43	66	611
Cd	ICP-QMS	µg/L	0.001	0.003	0.002	29	<0.003	0.0032	1.1
Ce	ICP-QMS	µg/L	0.0005	0.001	0.0001	13	<0.001	<0.001	6.2
Cl⁻	IC	mg/L	0.01	0.01	–	–	0.18	13	3,627
Co	ICP-QMS	µg/L	0.002	0.01	0.002	5	<0.01	0.023	16
Cr	ICP-QMS	µg/L	0.01	0.2	0.014	7	<0.2	<0.2	27
Cs	ICP-QMS	µg/L	0.0005	0.002	0.004[c]	3	<0.002	0.039	415
Cu	ICP-QMS	µg/L	0.01	0.1	0.029	2	<0.1	0.27	100
Dy	ICP-QMS	µg/L	0.0002	0.001	0.001	16	<0.001	0.0012	0.39
EC	conductometric	µS/cm	–	–	–	–	18	588	26500
Er	ICP-QMS	µg/L	0.0002	0.001	0.0004	13	<0.001	<0.001	0.77
Eu	ICP-QMS	µg/L	0.0002	0.001	0.001	18	<0.001	<0.001	0.45
F⁻	IC	mg/L	0.003	0.003	–	–	<0.003	0.19	11
Fe	ICP-QMS	µg/L	0.01	0.5	0.08	4	<0.5	0.69	13,500
Ga	ICP-QMS	µg/L	0.0005	0.005	0.005	4	<0.005	0.005	3.9
Gd	ICP-QMS	µg/L	0.0002	0.002	0.001	22	<0.002	<0.002	0.66
Ge	ICP-QMS	µg/L	0.005	0.03	0.02	6	<0.03	<0.03	110
Hf	ICP-QMS	µg/L	0.0001	0.002	0.001	28	<0.002	<0.002	1.6

(continued)

Table 10.1 (continued)

Parameter	Method	Unit	IDL*	RDL**	PDL***	Precision[a]/%	Min	Median	Max
Ho	ICP-QMS	µg/L	0.0001	0.001	0.0003	19	<0.001	<0.001	0.12
I	ICP-QMS	µg/L	0.01	0.2	0.09	15	0.2	4.8	4,030
K	ICP-OMS	mg/L	0.05	0.1	–	–	0.1	2.1	558
La	ICP-QMS	µg/L	0.0001	0.001	0.001	9	<0.001	0.0023	10
Li	ICP-QMS	µg/L	0.01	0.2	0.14	5	0.2	10	9,860
Lu	ICP-QMS	µg/L	0.00005	0.001	0.0002	16	<0.001	<0.001	0.41
Mg	ICP-OES	mg/L	0.005	0.001	–	–	<0.01	16	4,010
Mn	ICP-QMS	µg/L	0.005	0.1	0.111	2	<0.1	0.54	1,870
Mo	ICP-QMS	µg/L	0.001	0.02	0.017	4	<0.02	0.28	74
Na	ICP-OES	mg/L	0.02	0.1	–	–	0.4	16	8,160
Nb	ICP-QMS	µg/L	0.001	0.01	0.001	15	<0.01	<0.01	0.54
Nd	ICP-QMS	µg/L	0.0001	0.001	0.001	18	<0.001	0.0021	5.1
NH_4^+	Photometric	mg/L	0.005	0.005	–	–	0.0025	0.0025	60
Ni	ICP-QMS	µg/L	0.005	0.02	0.01	4	<0.02	0.18	95
NO_3^-	IC	mg/L	0.01	0.01-0.1[b]	–	–	<1	1.3	995
P	ICP-OES	µg/L	0.01	0.01	–	–	<6.5	33	2,863
Pb	ICP-QMS	µg/L	0.001	0.01	0.003	6	<0.01	0.016	2.3
pH	Potentiometric	–	–	–	–	–	4	6.8	9.9
Pr	ICP-QMS	µg/L	0.00005	0.001	0.0002	15	<0.001	<0.001	1.5
Rb	ICP-QMS	µg/L	0.001	0.01	0.30[c]	6	0.015	2.1	631
Sb	ICP-QMS	µg/L	0.001	0.01	0.0001	6	<0.01	0.27	4.4
Se	ICP-QMS	µg/L	0.005	0.02	0.009	19	<0.02	0.054	371
Si	ICP-OES	mg/L	0.03	0.03	–	–	0.42	6.5	59
Sm	ICP-QMS	µg/L	0.0001	0.001	0.001	23	<0.001	0.0013	0.67
Sn	ICP-QMS	µg/L	0.001	0.02	0.003	12	<0.02	<0.02	1.8
SO_4^{2-}	IC	mg/L	0.01	0.01	–	–	0.01	20	20,342
Sr	ICP-OES	µg/L	0.001	0.001	–	–	2	326	25,500
Ta	ICP-QMS	µg/L	0.0005	0.005	0.002	–[d]	<0.005	<0.005	0.037

	Method	Units	*	**	***	n			
tAlk	titration	mg/L	0.1	0.1	–	–	<2	286	16,110
tAlk	titration	meq/L	–	–	–	–	<0.03	4.7	264
Tb	ICP-QMS	µg/L	0.00005	0.001	0.0002	23	<0.001	<0.001	0.077
Te	ICP-QMS	µg/L	0.001	0.03	0.01	–(d)	<0.03	<0.03	0.32
Th	ICP-QMS	µg/L	0.0001	0.001	0.0004	33	0.0005	0.0005	0.15
Ti	ICP-QMS	µg/L	0.005	0.08	0.04	52	0.04	0.04	6.3
Tl	ICP-QMS	µg/L	0.0005	0.002	0.001	6	<0.002	0.0041	2.2
Tm	ICP-QMS	µg/L	0.00005	0.001	0.0002	22	<0.001	<0.001	0.19
U	ICP-QMS	µg/L	0.00005	0.001	0.006(c)	2	0,0005	0.23	229
V	ICP-QMS	µg/L	0.01	0.1	0.02	6	<0.1	0.17	49
W	ICP-QMS	µg/L	0.002	0.05	0.009	1	<0.05	<0.05	28
Y	ICP-QMS	µg/L	0.00005	0.001	0.002	7	<0.001	0.012	3.5
Yb	ICP-QMS	µg/L	0.0001	0.001	0.001	17	<0.001	<0.001	1.8
Zn	ICP-QMS	µg/L	0.01	0.2	0.11	3	<0.2	0.89	651
Zr	ICP-QMS	µg/L	0.0001	0.001	0.013	7	<0.001	0.0075	165

* Instrumental detection limit; ** reported detection limit; *** practical detection limit

(a) Precision at the 95% confidence level

(b) Depends on total dissolved solids (TDS)

(c) Too large concentration range and too many high values to reliably calculate PDL

(d) Insufficient values above detection limit to estimate precision

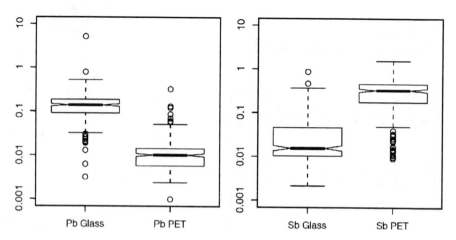

Fig. 10.1 Box plot comparison of the same 131 mineral water brands sold in both glass and PET bottles. Leaching of Pb from glass and of Sb from PET is clearly indicated. Analytical results in μg/L (from [25] Fig. 22, p.48)

materials to stored water does clearly occur. For example, with respect to PET bottles Sb was identified as the main problematic element, whereas for glass leaching Pb (and some additional elements) has been shown to be considerable in relation to natural concentrations of these elements in bottled water (Fig. 10.1).

Because it was practically impossible to buy only water in a certain bottle material and colour, it was necessary to test the influence of different bottle materials and colours on element concentrations observed in stored water [25, 28, 29]. The leaching test was carried out on 126 bottles filled with high purity water (18.2 MΩ/cm), and element concentrations were measured after 1, 2, 3, 4, 5, 15, 30, 56, 80 and 150 days. The leaching tests were performed at two different pH values (3.5 and 6.5), and results were directly compared to actual element concentrations in bottled water. The bottle leaching results are summarised below:

- In relation to natural variation observed for the various elements in bottled water, leaching from bottle materials to stored water is a minor problem at a pH of 6.5. At a pH of 3.5 it becomes, however, a serious problem for quite a number of elements (e.g., glass bottles: Al, As, B, Ba, Be, Bi, Ca, Cd, Ce, Co, Cr, Cs, Cu, Ga, Hf, I, K, La, Li, Mg, Mn, Mo, Na, Nb, Ni, Pb, Rb, Sb, Se, Sn, Sr, Ta, Te, Th, Ti, Tl, U, V, W, Zn; soft PET: Ag, Dy, Er, Fe, Gd, Ge, Ho, Lu, Nd, Pr, Sm, Tb, Tm, V, Y, Yb, Zr) [28]
- PET bottles contaminate the water with Sb to an extent where all measured values cannot any longer be used to investigate natural Sb concentration in the water.
- Glass bottles contaminate stored water with a considerably longer list of elements: Ce, Pb, Al, Zr, Ti, Hf, Th, La, Pr, Fe, Zn, Nd, Sn, Cr, Tb, Ag, Er, Gd, Bi, Sm, Y, Lu, Yb, Tm, Nb and Cu; some glass bottles also leach Sb.
- Green glass bottles leach more Cr (Fe, Zr) to stored water than clear glass bottles and, in general, dark coloured bottles (brown, green, blue, independent of material) leach more than clear bottles.

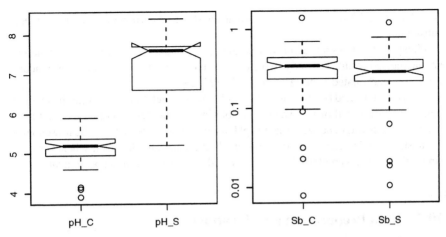

Fig. 10.2 Box plot comparison of pH and Sb concentration in carbonated (C) and still (S) waters, for the same 131 pairs of bottled mineral water that are marketed both in still and carbonated variants. While pH is strongly affected, an effect on the Sb concentration is hardly discernible and statistically insignificant. Analytical results of Sb in μg/L, and pH in units (from [25] Fig. 24, p.50)

For the purposes of the published geochemical atlas [25], the following conclusions were reached:

1. Bottled water cannot be used to establish the natural concentration range and variation of Sb, since results from PET bottles were preferentially used for mapping.
2. Ce, Cr, Pb and Al concentrations observed in bottled water can be seriously influenced by glass bottles.
3. The majority of elements can be used to produce geochemical maps that are not seriously influenced by contamination from bottle materials.

10.4.2   Carbonatisation Effects

Naturally carbonated waters do occur, but carbon dioxide is also added artificially to many still waters to obtain a sparkling "mineral water" variety. The content of most solutes in water should not be affected. However, for geochemical mapping, it was important to check the magnitude of any effects on the observed water chemistry for all parameters determined, because for completeness of coverage results of still and carbonated water were combined.

The major effects of adding CO_2 to bottled water is on pH and alkalinity. In total, 131 samples were obtained, where the same water was sold in a still and in a sparkling variant [25]. This subset of samples allowed the direct comparison of analytical results. The pH of carbonated bottled water is considerably lower than its corresponding still water (Fig. 10.2). The leaching tests have shown that a lower pH will increase leaching of elements from the bottle material to the stored water [28].

A difference of 2 pH units in the median of still and carbonated water is indeed substantial.

Phosphorus shows a much lower median in still than carbonated water, which is an unexpected result. The only two elements that otherwise indicate higher concentrations in carbonated water are Pb and Th [25, 28].

Both elements had been identified as problematic in relation to leaching from glass bottles, and most carbonated waters are sold in glass bottles. No other element is in fact affected. It was concluded that the pH of carbonated water may not be low enough to substantially increase leaching from the bottle material. However, whenever possible, still water was preferred over carbonated water for plotting geochemical maps.

10.5   Data Preparation and Treatment

Since, in many cases, several brands of bottled water come from the same locality, it was decided to reduce the data set to "one bottle per site" for presentation in the geochemical atlas [25]. European Union regulations require that a *bona fide* mineral water must have a single well dedicated to a given brand. Thus, at many mineral water bottling plants, it is common practice to have several wells, each one dedicated to a specific "brand" name. There will often be some real and significant hydrochemical variation between these different "brand" wells, even at a single site. Hence, by reducing the data set from 1,247 to 884 samples, i.e., "one bottle per site", some degree of actual within site hydrochemical variation is discarded.

Due to the results of the bottle leaching test [25, 28, 29], and the comparison between carbonated and non-carbonated water, a list of priorities was established for which bottle to choose as being representative for a site:

1. Non-carbonated (still water) was preferred over carbonated water;
2. Clear PET-bottles were preferred against all other bottle materials and colours, and
3. Samples sold in glass bottles were only left in the data set when no other bottle type was available.

The above procedure resulted in a final data set containing 884 samples that was used for the extraction of statistical parameters, and construction of graphs, plots and distribution maps [25]. Values below detection limit were set to half the detection limit for all graphics, including distribution maps and statistical calculations.

Exploratory data analysis techniques were used for statistical analysis [38]. All graphics and statistical calculations were prepared in R [30]. Analytical results expressed in mg/L (or µg/L) are compositional data [1], and the calculation of mean or standard deviation does not make sense for such data [16].

Mapping ground water geochemical data is plagued by many problems. The distribution of sample locations is usually very uneven, with large gaps on the maps where no samples could be taken – this is an undesirable situation in terms of obtaining a good impression of the regional distribution of investigated parameters [26]. It is, for example, not possible to produce smoothed colour surface maps from such data sets; even to plot readable point source maps is a challenge. Because the distribution of anomalously "high" values in Europe is probably the most interesting aspect of the published geochemical atlas [25], the 'variable-size dot' or 'growing dot' technique, as originally suggested by Björklund and Gustavsson [4], was used for producing all distribution maps. The technique, together with its advantages and disadvantages, is discussed in detail by Reimann et al. [30]. The variable-size dots grow exponentially according to element concentration between the 15th and 99th percentile [17, 30].

10.6   Results

The results of European ground water geochemistry (Table 10.1), using bottled water as a sampling medium, are described in the published geochemical atlas [25], in a special issue of the Journal of Exploration Geochemistry [2], and other publications [11, 21, 28, 29]. Here, only some key results are presented and concisely discussed.

10.6.1   Major Ion Geochemistry

Each ground water has a somewhat unique hydro chemical fingerprint that reflects the balance of all the various processes during its evolution, its residence time in the aquifer, the mineralogy of rocks and sediments that it comes into contact with, and so on. It is possible to plot major cations and anions as milliequivalent proportions on a so-called Durov diagram (Fig. 10.3). The major cations ($Na^+(+ K^+)$, Mg^{2+}, Ca^{2+}) are plotted on the top triangle, while the major anions (Cl^-, SO_4^{2-} and alkalinity or HCO_3^-) are plotted on the left triangle. The points are then projected onto a square central field. On this diagram, dilute, newly recharged waters, which may still possess a weak signature of marine salts in coastal areas, would plot as small dots (i.e., low electrical conductivity) in the lower right field – they would be termed as a "low ionic strength Na–Cl" water type. More "normal" ground waters, albeit still of relatively low mineralisation, of Ca–HCO$_3$ type, derived from calcite hydrolysis (Eq. 10.1), plot in the top left.

$$CO_2 \quad +H_2O \quad +CaCO_3 \quad \rightarrow \quad Ca^{2+} \quad +2HCO_3^-$$

Carbon dioxide   + Water   + Calcite   + Calcium   + bicarbonate(dissolved)   (10.1)

More evolved granitic ground waters, characterised by prolonged aluminosilicate weathering, may be of Na–HCO$_3$ type (Eq. 10.2), and plot at the top right. Deep saline brines would most likely be of Na–Cl composition and plot as large-diameter dots (high electrical conductivity) near the bottom right of the diagram.

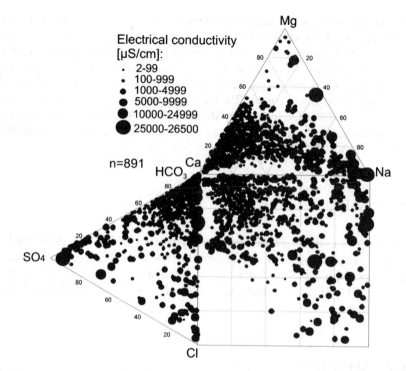

Fig. 10.3 Durov diagram for the European bottled water data set (N = 884). The diagram is based on milliequivalent fractions of the major cations and anions. The dot size is related to total dissolved solutes (on the basis of electrical conductivity). note that for plotting the Durov diagram only Na and not Na + K was used (from [25] Fig. 7, p.16)

$$2NaCaAl_3Si_5O_{16} \ +6CO_2 \ +9H_2O \ \rightarrow \ 2Ca^{2+} \qquad +2Na^+$$
$$+6HCO_3^- \qquad +3Al_2Si_2O_5(OH)_4$$
$$+4SiO_2$$

| feldspar | carbon dioxide | water | +calcium +bicarbonate +silica | +sodium +clay | (10.2) |

The Durov diagram shows the large variety of water types that are sold as bottled water in Europe, and describe the major ion geochemistry of ground water. The majority of samples fall, however, into the calcium bicarbonate corner of the diagram (Fig. 10.3).

10.6.2 Arsenic Distribution

Arsenic in bottled water varies from <0.03 to 89.8 µg/L (n = 884), with a median of 0.235 µg/L (Table 10.1; [25]), whereas in European surface water, As ranges from <0.01 to 27.3 µg/L (n = 807), and its median is slightly higher at 0.63 µg/L [33].

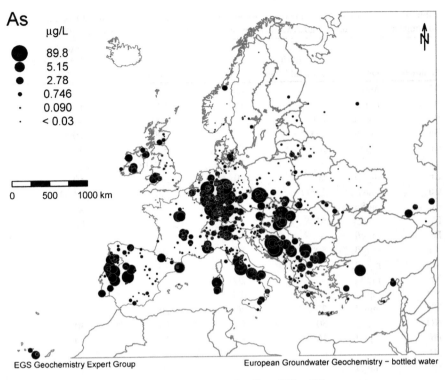

Fig. 10.4 Distribution of As in bottled water samples (from [26], p.77)

The As distribution map in bottled water shows a number of areas with enhanced values (Fig. 10.4). One of the interesting patterns, observed for As in the geochemical atlas of Europe [33], was generally the higher As concentrations in Southern than in Northern Europe (a factor 3 difference in the median value). A similar pattern can be seen for bottled water, As values are clearly higher in Southern European countries (e.g., median for Northern Europe = 0.07, and Southern Europe = 0.28 µg As/L). Many of the high As values, displayed on the map, occur in parts of Europe that are known for the occurrence of sulphide mineralisation, e.g., Portugal, Spain, Pyrenees, Massif Central, South Poland. Furthermore, the alkaline volcanic provinces in Italy are marked by higher than usual As values. The highest As value (89.8 µg/L) was reported in a Polish mineral water, abstracted in an area with known mineralisation. Even four samples from Germany are above the drinking water standards (Table 10.2); one is abstracting water from Palaeozoic rocks, and drawing water from a major fault zone. It is important to mention what is not seen on the map, i.e., in the Southern parts of the Great Hungarian Plain there occurs ground water with high natural concentrations of As [35], which are, of course, not bottled, but would add on another order of magnitude to the natural variation observed for As. Bedrock ground water with high As concentrations from Finland and Sweden are also not displayed in this low

Table 10.2 Statutory and guideline values according to EU Directives, WHO and FAO for mineral and drinking water (all values in μg/L, except Cl⁻, F⁻, Na, NH_4^+, NO_3^- and SO_4^{2-} in mg/L; EC μS/cm)

Parameter	Unit	[14]/40/EC Mineral Water	[13]/83/EC Drinking Water	WHO [40]	FAO Codex [15] (Revision 1997)
Al	μg/L	–	200[g]	100[Sm] & 200[L]	–
As	μg/L	10	10	10[P]	10
B	μg/L	(a)	1,000	500[T]	–
Ba	μg/L	1,000	–	700	700
Cd	μg/L	3	5	3	3
Cl⁻	mg/L	–	250[g]	<250	–
Cr	μg/L	50	50	50[P]	50
Cu	μg/L	1,000	2,000	2,000	1,000
EC	μS/cm	–	2,500[g]	–	–
F⁻	mg/L	5[b]	1.5	1.5	(c)
Fe	μg/L	–	200[g]	–	–
Mn	μg/L	500	50[g]	400[K]	500
Mo	μg/L	–	–	70	–
Na	mg/L	–	200[g]	–	–
NH_4^+	mg/L	–	0.5[g]	–	–
Ni	μg/L	20	20	20	20
NO_3^-	mg/L	50	50	50	50
Pb	μg/L	10	10	10	10
pH	–	–	≥6.5 – ≤9.5	–	–
Sb	μg/L	5	5	20	5
Se	μg/L	10	10	10	10
SO_4^{2-}	mg/L	–	250[g]	–	–
U	μg/L	–	–	15[P],[T]	–

(a) Pending

(b) Natural mineral waters with a fluoride concentration exceeding 1.5 mg/L shall bear on the label the words 'contains more than 1.5 mg/L of fluoride: not suitable for regular consumption by infants and children under 7 years of age'

(c) Where the water contains more than 1.5 mg/L fluoride, the label shall read "The product is not suitable for infants and children under the age of seven years"

(g) Guideline value

(K) Concentrations of the substance at or below the health-based guideline value may affect the appearance, taste or odour of the water, resulting in consumer complaints

(L) Large water treatment facilities

(Sm) Small water treatment facilities

(P) Provisional, as there is evidence of a hazard

(T) provisional, because calculated guideline value is below the level that can be achieved through practical treatment methods, source protection, etc.

sample density map. In general, the map may display lower As-concentrations than one would observe when collecting raw, untreated ground water. Even the few allowed water treatment options for mineral water (e.g., oxygenation) will lower the observed As-concentration.

Bottle leaching should have no influence on the observed As concentrations (up to 0.09 µg/L from glass bottles).

The water standard defined by European Union directives for As in mineral water and drinking water is 10 µg/L (Table 10.2), and nine samples exceed this level [7, 25].

10.6.3   Chromium Distribution

Chromium in bottled water varies from <0.2 to 27.2 µg/L (n = 884), with a median of <0.2 µg/L (Table 10.2; [25]), whereas in European surface water, Cr ranges from <0.01 to 43 µg/L (n = 806), and its median is slightly higher at 0.38 µg/L [33].

The main pattern on the Cr distribution map is a large anomaly in SE-Europe (Albania, Hellas) (Fig. 10.5). Northern Europe shows generally slightly lower Cr-levels in bottled water. Geology, and especially the occurrence of mafic and ultramafic rocks (such as ophiolites – remnants of oceanic crust thrust up onto land during mountain building episodes), is very important for the distribution of Cr in the surface environment. Over 2,000 mg/kg Cr occur in rocks and soils associated with the presence of ultramafic rocks. It is, thus, perhaps not surprising that the main anomaly on the distribution map is located in SE-Europe, with its abundance of ophiolites in Albania and Hellas. Note also the high value in Cyprus, with its famous Troodos ophiolite complex. Many enhanced Cr values are observed in Germany, where the use of glass bottles is widespread. One can speculate whether leaching of Cr from the glass bottles could lead to somewhat enhanced German Cr concentrations.

The six highest Cr values (up to 27.2 µg/L), however, are all reported from Hellas and are related to the occurrence of ultramafic rocks.

The bottle leaching test has shown that glass bottles (especially green glass) can leach significant amounts of Cr (up to 0.2 µg/L – 2 µg/L at pH 3.5) to the stored water.

The European Union limit for chromium in drinking water and in mineral water is 50 µg/L (Table 10.2; [14]). Even the maximum concentration found in European bottled water (27.2 µg/L) is well below this limit.

10.6.4   Fluoride Distribution

Fluoride in bottled water varies from 0.003 to 10.7 mg/L (n = 884), with a median of 0.188 mg/L (Table 10.1; [25]), whereas in European surface water, F⁻ ranges from <0.05 to 1.55 mg/L (n = 808), and its median is slightly lower at 0.1 mg/L [33].

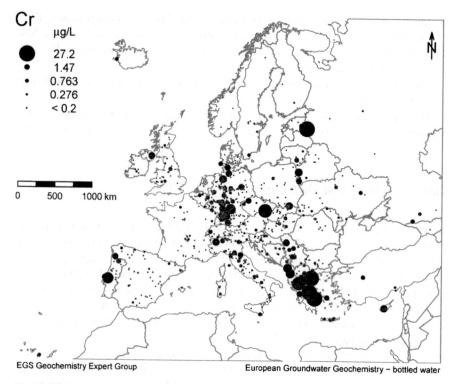

Fig. 10.5 Distribution of Cr in bottled water samples (from [28], p.99)

The distribution map of F⁻ in bottled water samples shows a predominance of high values in the waters from North-eastern Europe, where many "true" (i.e., highly mineralised) mineral waters are bottled (Fig. 10.6).

Several wells, abstracting water in areas underlain by young granite, exhibit unusually high F⁻ concentrations (e.g., Portugal, France). In Italy, bottled water extracted from wells in areas with Tertiary volcanic rocks show elevated F⁻ values. In Bulgaria, the high F⁻ concentrations are possibly due to Palaeozoic granite and fluoride mineralisation (e.g., Mihalkovo deposit). The highest F⁻ value of 10.7 mg/L was reported from a well in Georgia, and it was marked as "medicinal water". Many of the anomalous wells can be traced to deep fault zones.

Bottle leaching should have no influence on the observed F⁻ concentrations in bottled water, although anions were not tested.

The mineral water standard for F⁻ is 5 mg/L, while for drinking water the EU standard is 1.5 mg/L (Table 10.2).

Fluorine is an interesting element because the window of beneficial F⁻ concentrations in drinking water is very narrow (0.5–1.5 mg/L), compared to natural variation [12]. Too little F⁻ can weaken teeth's resistance to dental caries, while

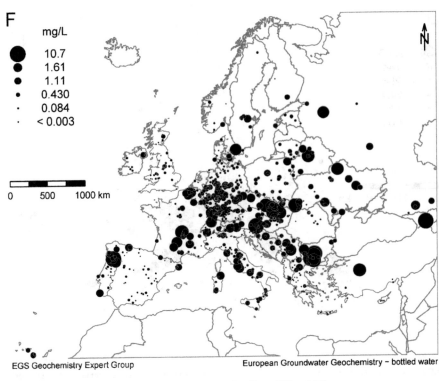

Fig. 10.6 Distribution of F⁻ in bottled water samples (from [28], p.111)

prolonged consumption of too much F⁻ can lead to dental and finally (>5 mg/L) skeletal fluorosis. Some authors have speculated that a limit of 1.5 mg F⁻/L may even be too high in warm countries where much water is drunk.

Only one sample in the bottled water data set exceeds the EU limit of 5 mg F⁻/L, but more than 5% of all bottled waters would fail the more stringent drinking water standard of 1.5 mg F⁻/L (Table 10.2). The reasoning of the regulators is probably that nobody drinks mineral water in large quantities on a daily basis – although in the light of current market trends, such an assumption may be dangerous and could lead to certain people being overexposed to fluoride through the "mineral water" pathway.

10.6.5   Lithium Distribution

Lithium in bottled water varies from <0.2 to 9,860 μg/L (n = 884), with a median of 10.0 μg/L (Table 10.1; [28]), whereas in European surface water, Li ranges from <0.005 to 356 μg/L (n = 808), and its median is lower at 2.10 μg/L [33].

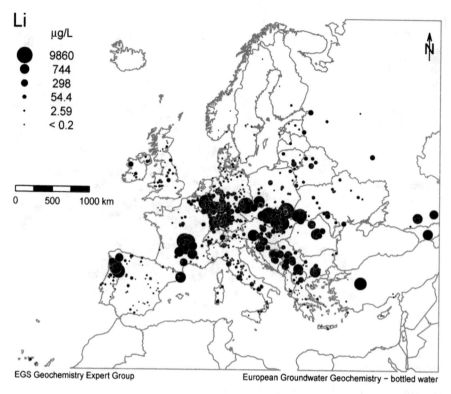

Fig. 10.7 Distribution of Li in bottled water samples (from [28], p.131)

The distribution map depicts a number of distinct Li anomalies (Fig. 10.7). Many of the 'hot spots' (e.g., Northern Portugal, France) are related to young granitic intrusions; similar patterns are displayed by K, Cs and Rb, elements associated with felsic intrusives (see [25]).

Many wells from the general area of the Carpathian Mountains and the Dinarides also exhibit enhanced Li concentrations. In Germany, high Li concentrations are associated with wells in Jurassic and Triassic sediments, but a well drawing water from an important fault zone is also indicated by an unusually high Li value. The highest Li value (9,860 µg/L) was observed in bottled water from Slovakia. Lithium is one of the elements, together with B, Br, Cs, K, Ge and Rb, where high concentrations are rather characteristic of "mineral water".

Bottle leaching has no influence on the observed Li distribution on the map at circum-neutral pH, while minor leaching may occur from glass bottles at low pH (up to 3.23 µg Li/L).

No water standard is defined in the EU for Li. However, Li is a biologically active element and Li-based drugs have been used to treat manic-depressive condi-

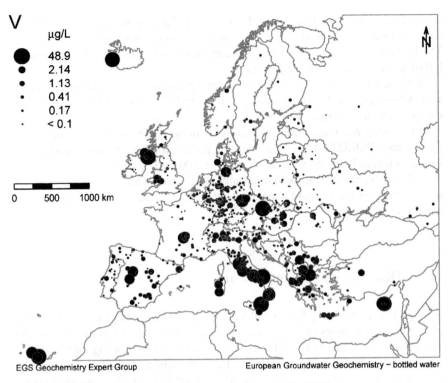

Fig. 10.8 Distribution of V in bottled water samples (from [28], p.189)

tions since the 1950s. It is thus quite likely that some of the bottled waters with the highest Li concentrations have medical effects. Given the enormous natural variation of Li (6 orders of magnitude), it is one of the elements where deficiency may play a role as well. According to Bradford [5], Li can be toxic to some plants at concentrations >60 µg/L.

10.6.6   Vanadium Distribution

Vanadium in bottled water varies from <0.1 to 48.9 µg/L (n = 884), with a median of 0.17 µg/L (Table 10.1; [25]), whereas in European surface water, V ranges from <0.05 to 19.5 µg/L (n = 807), and its median is slightly higher at 0.46 µg/L [33].

On the V distribution map (Fig. 10.8), all the active volcanic areas in Europe are clearly marked by anomalies (e.g., Iceland, Canary Islands, Cyprus, and Italy). In France, a V anomaly coincides with the Massif Central, possibly linked to volcanic lithologies. In North Ireland, the influence of the Palaeogene basalt is visible. In Northern Hellas, the V anomalies are associated with ophiolites, and possibly bauxite. The highest value (48.9 µg/L) was reported from an Italian bottled water and is linked to the alkaline volcanic province.

Bottle leaching can have an impact on the observed V concentrations of water sold in both, glass and PET bottles (up to 0.06 µg/L). At low pH bottle leaching is increased (up to 0.08 µg/L V).

No drinking water standard is defined for V in the EU. Bosnia and Herzegovina, Croatia and F.Y.R.O.M. have set a limit of 5 µg/L V for drinking water, and Serbia 1 µg/L, whereas Ukraine has defined a limit of 100 µg/L V [25].

Although vanadium can be biologically active, and is an essential nutrient for several micro-organisms and animals, its role in the human body is unclear and remains controversial.

10.7 Discussion and Conclusions

Geology is one of the key factors influencing the observed element concentrations in bottled water samples for a significant number of elements. Examples include the high values of Cr, clearly related to ophiolites; Li (Be, Cs) showing high values in areas underlain by Hercynian granites; F (K, Si) is connected with the occurrence of alkaline rocks, especially near the volcanic centres in Italy, and V indicates the presence of active volcanism. Elevated As values are apparently associated with sulphide mineralisation.

Some elements, as observed in bottled water, are clearly not representative for "normal" shallow groundwater, but tend to exhibit unusually high concentrations, typical for "mineral water", e.g., F and Li, but also B, Cs, Ge, Na, Rb, Te, Tl and Zr.

The data presented in the published geochemical atlas [25] can be used to gain a first impression of the natural variation of analysed elements in ground water at a European scale. Natural variation is enormous, usually spanning three to four and occasionally up to seven orders of magnitude.

Several elements, where no potable water standards are defined in Europe, show surprisingly high concentrations in bottled water. In terms of health effects, more attention at both ends of the concentration range (deficiency as well as toxicity) may be required for quite a number of elements.

For some elements the reported concentrations can be influenced by bottle material. However, only for Sb was bottle leaching, in comparison to its natural concentrations in water, so serious that the results could not be used to plot a distribution map or taken to represent its natural concentration and variation in Europe. In general, glass bottles leach more elements to the stored water than PET bottles. However, all values observed during the leaching tests were well below all maximum admissible concentrations (MAC) as defined for drinking water in Europe, and it can thus be tentatively concluded that bottle leaching is unlikely to represent a health risk. The bottle leaching test demonstrated that there exist bottles that do not leach any of the indicated elements to the stored water.

In terms of water standards, the majority of samples fulfil the requirements of the European Union legislation for mineral (and drinking) water. For some elements, a few bottled water samples exceed the potable water standards see, e.g., the maximum values observed for Al, As, Ba, F^-, Mn, Ni, NO_2^-, NO_3^-, Se and U. It must be noted that the maximum admissible concentration for F^- in bottled mineral water is set very high (5 mg/L instead of the 1.5 mg/L valid for drinking water) in order to avoid too many compliance failures, i.e., about 5% of all mineral water samples report F^- concentrations above 1.5 mg/L. This statutory practice is questionable in view of bottled water increasingly replacing tap water as general drinking water. With very few exceptions (Al, As, B, Ba, Fe, Mn, Ni, and Se – [7]), all values reported in the study of European ground water geochemistry, using bottled water as sampling medium are, however, well below the MAC values as defined by European legislation.

There exist a number of elements (Be, I, Li, Tl and U) that have been indicated for having health effects in international literature, but for which no MAC values are defined in the European Union. Some of these elements exhibit a very large natural variation in bottled water.

It can be concluded that the idea of using bottled water as a first proxy for ground water geochemistry and quality at the European scale was not as absurd as it might, at first glance, have appeared. Despite all potential problems, it has been shown that natural variation in ground water quality is much larger than the impact of any secondary consideration. Thus, on most element distribution maps, the importance of geology and other natural processes on the chemical composition of ground water is clearly visible.

10.8   Acknowledgements

Without the dedicated input of all members of the EuroGeoSurveys Geochemistry EGG Team and associated friends and colleagues this project would have not been possible. Thanks are due to Adriana Ion (Geological Institute of Romania); Aivars Gilucis (Latvian Environment, Geology and Meteorology Agency); Benedetto De Vivo, Annamaria Lima and Stefano Albanese (University of Napoli Federico II, Italy); Bjørn Frengstad, Rolf Tore Ottesen and Ola Eggen (Geological Survey of Norway); Børge Johannes Wigum (Mannvit hf, Iceland); Boris I. Malyuk, Volodymyr Klos and Maryna Vladymyrova, (Geological Survey of Ukraine); Carla Lourenço and Maria Joao Batista (Laboratório Nacional de Energia e Geologia, Portugal); Corina Ionesco (University of Cluj-Napoca, Romania); David Banks (Holymoor Consultancy Ltd., U.K.); Dee Flight, Shaun Reeder and Pauline Smedley (British Geological Survey); Domenico Cicchella

(University of Sannio, Italy); Edith Haslinger (Austrian Institute of Technology); Enrico Dinelli (University of Bologna, Italy); Friedrich Koller (University of Vienna); Gerhard Hobiger and Albert Schedl (Geological Survey of Austria); Gyozo Jordan, Ubul Fugedi and Laszlo Kuti (Geological Institute of Hungary); Hazim Hrvatovic, Neven Miosic, Ferid Skopljak and Natalija Samardzic (Geological Survey of the Federation of Bosnia and Herzegovina); Ignace Salpeteur and Christophe Innocent (BRGM, France); Ilse Schoeters (Rio Tinto Minerals); Jaan Kivisilla and Valter Petersell (Geological Survey of Estonia); Josip Halamić and Ajka Šorša (Croatian Geological Survey); Juan Locutura, Alejandro Bel-lan and Mar Corral (Instituto Geologico y Minero de Espana); Kaj Lax and Madelen Andersson (Geological Survey of Sweden); Kujtim Onuzi (Institute of Geoscience, Technical University of Tirana); Lech Smietanski (Polish Geological Institute); Liida Bityukova (Institute of Geology at Tallinn University of Technology, Estonia); Maria Titovet (Ministry for Ecology and Nature of Republic Moldova, Department of Hydrogeology); Marianthi Stefouli, Panagiotis Sampatakakis, Constantinos Nikas, Athanasios Hatzikirkou, Eleftheria Poyiadji, Natalia Spanou, Georgios Vrachatis, Dimitris Dimitriou, Michalis Vavrados, Evangelos Nicolaou and Pavlos Vekios (Institute of Geology and Mineral Exploration, Hellas); Mateja Gosar (Geological Survey of Slovenia); Milena Zlokolica-Mandic, Tanja Petrovic and Aleksandra Gulan (Geological Institute of Serbia); Miloslav Duris (Czech Geological Survey); Nebojsa Veljkovic (Serbian Environmental Protection Agency); Neda Devic (Geological Survey of Montenegro); Nikolay Phillipov (SC Mineral, Russia); Olga Karnachuk (Tomsk State University, Russia); Paolo Valera (University of Cagliari, Italy); Peter Filzmoser (Institute of Statistics and Probability Theory, Vienna University of Technology); Peter Hayoz (Federal Office of Topography, Swiss Geological Survey); Peter Malik (State Geological Institute of Dionyz Stur, Slovakia); Raymond Flynn (Queen's University Belfast, Northern Ireland); Reijo Salminen, Timo Tarvainen and Jaana Jarva (Geological Survey of Finland); Robert Maquil (Service Géologique du Luxembourg); Trajce Stafilov (Sts. Cyril and Methodius University, Skopje, F.Y.R.O.M.), Uwe Rauch, Lars Kaste, Hans Lorenz and Fabian Jähne (Federal Institute for Geosciences and Natural Resources, Germany); Valeri Trendavilov (Department of Geology and Permits for Exploration, Subsurface and Underground Resources Office, Ministry of Environment and Water, Bulgaria); Virgilija Gregorauskiene (Lithuanian Geological Survey), and Walter De Vos (Geological Survey of Belgium). The members of the laboratory staff of BGR (Hans Lorenz, Wolfgang Glatte, Bodo Harazim, Fred Flohr, Anna Degtjarev, Jürgen Rausch) are thanked for their dedicated work on the analysis of the bottled water samples. Finally, we would like to thank the Director of NGU, Dr. Morten Smelror, the President of BGR, Professor Hans-Joachim Kümpel, and the EuroGeoSurveys Secretary General, Dr. Luca Demicheli, for their support of the project.

References

1. Aitchison J (1986) The statistical analysis of compositional data. Blackburn Press, Caldwell, 416 pp
2. Birke M, Demetriades A, De Vivo B (eds) (2010a) Mineral waters of Europe. Special Issue, J Geochem Exploration 107(3):217–422
3. Birke M, Reimann C, Demetriades A, Rauch U, Lorenz H, Harazim B, Glatte W (2010b) Determination of major and trace elements in European bottled mineral water—analytical methods. In: Birke M, Demetriades A, De Vivo B (eds) Mineral waters of Europe. Special Issue, J Geochem Exploration 107(3):217–226
4. Björklund A, Gustavsson N (1987) Visualization of geochemical data on maps: New options. J Geochem Explor 29:89–103
5. Bradford GR (1963) Lithium survey of California's water resources. Soil Sci 96:77–81
6. Demetriades A (2009) Quality control procedures in applied geochemical surveys. Open File Report. Institute of Geology and Mineral Exploration, Hellas, Athens, In Greek with an English summary
7. Demetriades A (2010) Use of measurement uncertainty in a probabilistic scheme to assess compliance of bottled water with drinking water standards. In: Birke M, Demetriades A, De Vivo B (Guest eds) Mineral waters of Europe. Special Issue, J Geochem Exploration 107(3): 410–422
8. Demetriades A (2011) Understanding the quality of chemical data from the urban environment – Part 2: Measurement uncertainty in the decision-making process. In: Johnson CC, Demetriades A, Locutura J, Ottesen RT (eds) Mapping the chemical environment of urban areas. Wiley-Blackwell, Chichester, pp 77–98
9. Demetriades A, Karamanos E (2003) Quality assurance and quality control (QA/QC) for in-situ geochemical methods, estimation of measurement uncertainty and construction of probability risk assessment maps. Network oriented risk assessment by In-situ screening of contaminated sites (NORISC), European Commission co-financed project, EVK4-CT-2000-00026 NORISC consortium report, Cologne, 20 pp
10. De Vos W, Tarvainen T (Chief eds), Salminen R, Reeder S, De Vivo B, Demetriades A, Pirc S, Batista MJ, Marsina K, Ottesen RT, O'Connor PJ, Bidovec M, Lima A, Siewers U, Smith B, Taylor H, Shaw R, Salpeteur I, Gregorauskiene V, Halamic J, Slaninka I, Lax K, Gravesen P, Birke M, Breward N, Ander EL, Jordan G, Duris M, Klein P, Locutura J, Bel-lan A, Pasieczna A, Lis J, Mazreku A, Gilucis A, Heitzmann P, Klaver G, and Petersell V (2006) Geochemical atlas of Europe. Part 2 – Interpretation of geochemical maps, Additional tables, figures, maps and related publications. Geological Survey of Finland, Espoo, Finland, 692 pp. Available online at: http://www.gtk.fi/publ/foregsatlas/. Accessed on 17 April 2011
11. De Vivo B, Birke M, Cicchella D, Giaccio L, Dinelli E, Lima A, Albanese S, Valera P (2010) Acqua di casa nostra. Le Scienze 508:76–85
12. Edmunds WM, Smedley P (2005) Fluoride in natural waters. In: Selinus O, Alloway B, Centeno JA, Finkelman RB, Fuge R, Lindh U, Smedley P (eds), Essentials of Medical Geology. Elsevier, Amsterdam, pp 301–329
13. EU directive 98/83/EC of 3rd Nov 1998 on the quality of water intended for human consumption. Official Journal of the European Communties, 05/12/1998, L330/32–54
14. EU directive 2003/40/EC/16-5-2003/ establishing the list, concentration limits and labelling requirements for the constituents of natural mineral waters and the conditions for using ozone-enriched air for the treatment of natural mineral waters and spring waters. Official Journal of the European Union, 22/5/2003, L126/34–39
15. FAO (1997) Codex standard for natural mineral waters. Codes Stan 108–1981, 5 pp
16. Filzmoser P, Hron K, Reimann C (2009) Univariate statistical analysis of environmental (compositional) data – Problems and possibilities. Sci Total Environ 407:6100–6108

17. Gustavsson N, Lampio E, Tarvainen T (1997) Visualization of geochemical data on maps at the Geological Survey of Finland. J Geochem Explor 59:197–207

18. Hall GEM (1998) Relative contamination levels observed in different types of bottles used to collect water samples. Explore 101(1):3–7

19. Keresztes S, Tatar E, Mihucz VG, Virag I, Majdik C, Zaray G (2009) Leaching of antimony from polyethylene terephthalate (PET) bottles into mineral water. Sci Total Environ 407: 4731–4735

20. Krachler M, Shotyk W (2009) Trace and ultra trace metals in bottled waters: Survey of sources worldwide and comparison with refillable metal bottles. Sci Total Environ 407: 1089–1096

21. Lima A, Cicchella D, Giaccio L, Dinelli E, Albanese S, Valera P, De Vivo B (2010) Che acqua beviamo. Le Scienze 501:68–77

22. Lloyd JW, Heathcote JA (1985) Natural inorganic hydrochemistry in relation to groundwater. Oxford Scientific Publications, Oxford University Press, 296 pp

23. Misund A, Frengstad B, Siewers U, Reimann C (1999) Natural variation of 66 elements in European mineral waters. Sci Total Environ 243(244):21–41

24. Nriagu JO, Lawson G, Wong HKT, Azcue JM (1993) A protocol for minimizing contamination in the analysis of trace metals in Great Lakes waters. J Great Lakes Res 19: 175–182

25. Reimann C, Birke M (eds) (2010) Geochemistry of European bottled water. Borntraeger Science Publishers, Stuttgart, 268 pp. Available online at: http://www.schweizerbart.de/ publications/detail/artno/001201002. Accessed on 17 Apr 2011

26. Reimann C, Garrett RG (2005) Geochemical background – concept and reality. Sci Total Environ 350:12–27

27. Reimann C, Wurzer F (1986) Monitoring accuracy and precision – improvements by introducing robust and resistant statistics. Mikrochimic Acta II, 1–6:31–42

28. Reimann C, Birke M, Filzmoser P (2010) Bottled drinking water: Water contamination from bottle materials (glass, hard PET, soft PET), the influence of colour and acidification. Appl Geochem 25(7):1030–1046

29. Reimann C, Birke M, Filzmoser P (2010) Reply to the comment "Bottled drinking water: Water contamination from bottle materials (glass, hard PET, soft PET), the influence of colour and acidification" by Hayo Müller-Simon. Appl Geochem 25(9):1464–1465

30. Reimann C, Filzmoser P, Garrett RG, Dutter R (2008) Statistical data analysis explained – applied environmental statistics with R. Wiley, Chichester, 343 pp

31. Reimann C, Grimstvedt A, Frengstad B, Finne TE (2007) White HDPE bottles as a source of serious contamination of water samples with Ba and Zn. Sci Total Environ 374: 292–296

32. Ross HB (1984) Methodology for the collection and analysis of trace metals in atmospheric precipitation. University of Stockholm, Department of Meteorology Report CM-67

33. Salminen R (Chief-ed), Batista MJ, Bidovec M Demetriades A, De Vivo B, De Vos W, Duris M, Gilucis A, Gregorauskiene V, Halamic J, Heitzmann P, Lima A, Jordan G, Klaver G, Klein P, Lis J, Locutura J, Marsina K, Mazreku A, O'Connor PJ, Olsson SÅ, Ottesen RT, Petersell V, Plant JA, Reeder S, Salpeteur I, Sandström H, Siewers U, Steenfelt A and Tarvainen T (2005) Geochemical atlas of Europe. Part 1 – Background information, methodology and maps. Geological Survey of Finland, Espoo, Finland, 526 pp. Available online at: http://www.gtk.fi/ publ/foregsatlas/. Accessed on 17 Apr 2011

34. Shotyk W, Krachler M (2007) Contamination of bottled waters with antimony leaching from polyethylene Terephthalate (PET) increases upon storage. Environ Sci Technol 41: 1560–1563

35. Shotyk W, Krachler M (2007) Lead in bottled water: contamination from glass and comparison with pristine groundwater. Environ Sci Technol 41:3508–3513

36. Shotyk W, Krachler M, Chen B (2006) Contamination of Canadian and European bottled waters with antimony leaching from PET containers. J Environ Monitor 8:288–292

37. Smedley PL, Kinniburgh DG (2002) A review of the source, behaviour and distribution of arsenic in natural waters. Appl Geochem 17:517–568
38. Tukey JW (1977) Exploratory data analysis. Addison Wesley, Reading, 688 pp
39. Westerhoff P, Prapaipong P, Shock E, Hillaireau A (2008) Antimony leaching from polyethylene terephthalate (PET) plastic used for bottled drinking water. Water Res 42: 551–556
40. WHO (2008) Drinking-water quality, 3rd edn incorporating the 1st and 2nd addenda. vol 1: Recommendations. World Health Organisation, Geneva, 668 pp. Available online at: http://www.who.int/water_sanitation_health/dwq/guidelines/en. Accessed on 17 Apr 2011

Part II
Remediation

Chapter 11
Advances in Bioremediation of Aquifers

Robert J. Steffan, Ph.D. and Guy W. Sewell, Ph.D.

Abstract Chlorinated solvents, including chlorinated ethenes and chlorinated ethane's, are primary contaminants of ground waters throughout the world. Because of their abundance, toxicity and chemical properties, treatment technologies have had to evolve beyond simple bio stimulation approaches, to effectively remediate chlorinated solvent contamination in a wide range of aquifer types. During the last 15 years a number of biological treatment approaches that allow cost-effective remediation have been developed, and new technologies continue to be developed to further improve treatment and to address situations where available technologies are not optimum. In this paper we discuss the most common active biological treatment technologies for chlorinated solvent remediation, including both bio stimulation and bio augmentation based approaches, and introduce a novel treatment technology, proton reduction, that is being developed for low cost and low maintenance remediation of chlorinated solvent-contaminated aquifers.

11.1 Introduction

Chlorinated ethenes and ethanes have been used extensively as industrial solvents and cleaning agents over the last 50 years. Their widespread use, historical and improper disposal practices, and the chemical properties and stability of the

R.J. Steffan, Ph.D. (✉)
Shaw Environmental, Inc, 17 Princess Rd, Lawrenceville, NJ 08648, USA
e-mail: rob.steffan@shawgrp.com

G.W. Sewell, Ph.D.
East Central University, 14th St, 1100 E, P-MB S-78 Ada, OK 74820, USA

F.F. Quercia and D. Vidojevic (eds.), *Clean Soil and Safe Water*, NATO Science for Peace and Security Series C: Environmental Security, DOI 10.1007/978-94-007-2240-8_11,
© Springer Science+Business Media B.V. 2012

compounds have resulted in their becoming common groundwater contaminants at many facilities throughout the world. As the result of the frequent occurrence of chlorinated solvent contamination, a number of treatment technologies have emerged and evolved to restore contaminated aquifers and protect drinking water supplies. These include pump and treat technologies that rely on extraction of the contaminated groundwater coupled with ex situ treatment (include air stripping, carbon adsorption, and chemical oxidation) to remove the contaminants. However this approach has proven to require high energy input and extremely long treatment times (years to decades), due to the adsorption of chlorinated solvents to aquifer materials. Air sparging technologies also have been developed to physically strip the contaminants from groundwater in situ, and this technology often is coupled with vapour extraction approaches for removing the stripped contaminants from the sub-surface. Again, this technology is energy intensive and it often is not suitable for treating deep aquifers or aquifers with highly heterogeneous or low permeability.

Currently, the most common treatment alternative for these contaminants involves reductive anaerobic biological in-situ treatment. Biodegradation can be performed by native organisms that use endogenous resources to support contaminant degradation (i.e., intrinsic bioremediation), or nutrients that are purposefully added to support their activity (i.e., bio stimulation). Biodegradation also can be performed by exogenous bacteria that are added to supplement the native microbial population (i.e., bio augmentation).

The predominant biodegradation pathway for chlorinated ethenes under anaerobic conditions is via bacterial-mediated reductive dechlorination during which chlorine atoms on the molecules are replaced by hydrogen. Because some bacteria use the chloroethenes as their primary electron acceptor, this process is often referred to as dehalorespiration. Sequential dechlorination of PCE typically proceeds to TCE, cis-1, 2-dichloroethene (cDCE), vinyl chloride (VC), and finally the desired end product, ethene. Although a wide range of bacteria are able to partially dehalogenate PCE and TCE, complete dehalogenation of chlorinated ethenes in the field has been linked to the presence of specific microorganisms of the genus *Dehalococcoides*, (a.k.a., DHC) [9, 10]. Hendrickson et al. [3], for example conducted a survey of multiple chlorinated ethene contaminated sites using a 16S rRNA gene molecular detection method. The results indicated that complete reductive dechlorination of chlorinated ethenes in situ strongly correlates with the presence of DHC and DHC-like bacteria. The sites lacking this microorganism exhibited incomplete dechlorination of PCE and TCE, and often had an accumulation of cDCE and VC. These microbes have now been extensively studied and their physiology is becoming well understood. Notably, the organisms appear to require chlorinated compounds as obligate electron acceptors, and the obligate electron donor appears to be H_2. In the strains studies to date, acetate serves as a carbon source for growth of the organisms [2]. If a particular strain of DHC is missing in an environment, degradation daughter products like cDCE and VC can accumulate and create increased health risks. In aquifers where cDCE and VC accumulate, bio augmentation often is considered and applied to ensure complete degradation of the target contaminants to non-toxic products.

11.2 Bio Stimulation

Bio stimulation (Fig. 11.1) involves adding a limiting nutrient to enhance the growth and activity of native microorganisms to stimulate the remediation of the target contaminant. The material added could be inorganic nutrients (e.g., nitrogen and phosphorous), but in the case of chlorinated solvent dehalogenation the limiting nutrient is often labile carbon that can serve as a source of reducing equivalents (electron donor) for reduction of the contaminants. Thus, for chlorinated solvent biodegradation, bio stimulation typically involves the addition of a fermentable carbon material, and because DHC require H_2 as an electron donor, the carbon source must be fermented to produce H_2 to stimulate dehalogenation. Many electron donor substrates have been successfully used for stimulating chlorinated solvent degradation, including soluble substrates (e.g., lactate, molasses, whey, sugars) and insoluble substrates (vegetable oils, poly-lactates, mulch, plant material, chitin). The selection of appropriate substrates typically depends on,

1. site hydrogeology;
2. ease of implementation;
3. treatment goals; and,
4. cost.

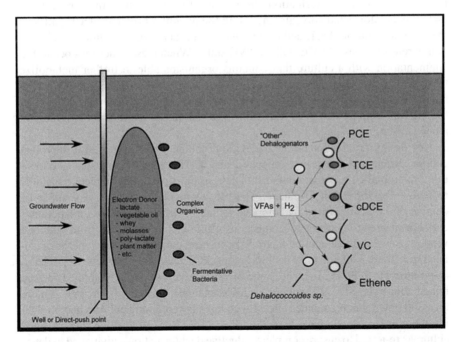

Fig. 11.1 Diagram of in situ bio stimulation for chlorinated solvent bioremediation. In a typical application a carbon source is injected into a contaminated aquifer where it is fermented by native bacteria to H_2 and volatile fatty acids (VFAs). The H_2 then serves as an electron donor for bacteria, including DHC that use the chlorinated solvent contaminants and their daughter products as respiratory electron acceptors

Generally, soluble substrates are more easily distributed through the aquifer, but they also are more rapidly utilized so they must be re-applied more frequently. Insoluble substrates are generally more difficult to distribute, although emulsification sometimes improves distribution, but they are slowly utilized and thus can be re-applied less frequently than soluble substrates. Insoluble substrates are often referred to as "slow-release" electron donors.

Bio stimulation is likely the most commonly used approach for remediating chlorinated solvent contaminated aquifers in the United States, although such statistics are not available, and many examples of this remedial approach can be readily found in technical literature.

11.3 Bio Augmentation

Reductive dechlorination, even by DHC, may occur by either of two distinct processes; cometabolic reductive dechlorination [9, 10] or dehalorespiration [1, 2]. Cometabolic reductive dechlorination of chlorinated ethenes is a relatively slow process and it can lead to the accumulation of cDCE and VC which are more toxic than the parent compounds. Dehalorespiration is usually a much more rapid form of reductive dechlorination. In aquifers that lack DHC that are able to metabolize chlorinated solvents, including the daughter products cDCE and VC, the accumulation of cDCE and/or VC may occur, and these conditions are commonly referred to as a "DCE stall" or "VC stall". When these conditions occur, bio augmentation with a culture that contains organisms able to further metabolize these compounds is often considered. Likewise, if rapid remediation of a site is required, bio augmentation may be a prudent remedial approach because it can limit the lag time required to grow sufficient DHC biomass in situ if bio stimulation alone is used.

Key design criteria for bio augmentation include identification of a microbial culture, enrichment and growth, injection, and distribution. The first step is to identify a microbial culture that contains a DHC strain capable of complete metabolic reductive dechlorination to ethene. The bio augmentation culture can either be obtained from a site exhibiting complete reductive dechlorination via a laboratory enrichment process, or an exogenous consortium can be identified from qualified vendors. Several stable, natural microbial consortia containing DHC have been isolated that are capable of fully dechlorinating PCE and TCE to ethene via dehalorespiration [3, 5, 8, 15, 16], and some of these have been widely applied for site remediation. The SDC-9™ culture [16], for example, has now been applied more than 300 times at contaminated sites. Procedures for enriching cultures and producing them at large scale have been described [16]. After production, the enriched culture is re-tested to ensure complete reductive dechlorination activity and to determine the DHC cell density, and it is then shipped to the site. At the site, the bio augmentation culture is injected into the subsurface via injection wells or direct push injection points. Distribution of the bio augmented culture is achieved using either groundwater recirculation or ambient groundwater flow. A carbon source is typically added prior to bio augmentation, or with the bio augmentation culture, in order to promote and maintain the highly reducing anaerobic conditions and to supply the carbon/electron donor supply needed for in situ growth of DHC and degradation of chlorinated ethenes.

The amount of microorganisms added to treat a site directly affects both the cost and performance of a remedial activity. The traditional approach for determining the number of organisms to add relied on calculating the approximate volume of groundwater to be treated (volume of the area to be treated multiplied by the porosity), and then adding enough culture to achieve approximately 1×10^7 DHC/L of the water to be treated. This estimation was based on early microcosm testing, but it was also supported by field evidence that suggested that this DHC density resulted in sufficient rates of biodegradation under field conditions [7]. The amount of microorganisms needed, however, actually also depends upon contaminant concentrations, site hydro geochemical conditions, competition by indigenous microorganisms, the relative concentration of DHC in the bio augmentation culture, in situ growth, transport, and decay of the bio augmented culture, and various other site-specific factors. Consequently, a more comprehensive analysis of cell dosage led to the development of an application model to more accurately evaluate cell dosage impacts [13, 15]. The use of the model, however, may be impractical for small sites where the incremental cost of bio augmentation culture relative to total remedial costs is small, especially given the success of the traditional volume-based estimate approach. In general, the cost of a bio augmentation culture for a given site constitutes only a few percent of the total remedial costs.

The main advantages of anaerobic bio augmentation with DHC are (1) complete reductive dechlorination of chlorinated ethenes to the innocuous by-product ethene, (2) reduced cleanup times, and (3) cost-effective remediation. One potential limitation to bio augmentation is that effective treatment is contingent upon adequate distribution of the degradative bacteria within the treatment area. Before implementing bio augmentation, or any in situ technology, an evaluation is necessary to consider site-specific characteristics and to determine the most effective treatment technology based on current contaminant and hydrogeochemical conditions and site access. A second potential limitation for successful bio augmentation is that unfavourable aquifer conditions such as low pH [16], low temperatures, elevated dissolved oxygen (DO) levels, or lack of adequate organic carbon may inhibit the activity of the bio augmentation culture or necessitate additional treatments like pH adjustment or pre-treatment to reduce DO levels. In addition, excessively low concentrations of chlorinated ethenes may not provide a sufficient source of electron acceptors needed to support dehalorespiration, thereby limiting in situ growth of the added culture. Excessively high concentrations of chlorinated ethenes may have a toxic effect on the added DHC population, but even aquifers with free product levels of solvents (i.e., DNAPL) and complex matrices can be successfully treated with bio augmentation [14].

11.4   Proton Reduction

In addition to fermentation of complex carbon, H_2 can be produced in water by electrolysis (electrochemically splitting water to H_2 and O_2 at opposing electrodes) or by the reduction of hydrogen ions (a.k.a., protons) through the introduction of an electrical current on certain metal surfaces (Fig. 11.2). This process is most commonly associated with electrodes, where reductive processes, including proton (H^+) reduction, occur on the surface of the cathode (i.e., negative electrode) at a relatively low potential (-0.5 V). The resulting cathodic H_2 has been demonstrated to support in situ biodegradation of chlorinated solvents [6, 17], and it also can support the production of acetate by homoacetogenic bacteria (either native or added as components of bio augmentation consortia).

This "Proton Reduction Technology" as applied by us was initially developed at the USEPA Robert S. Kerr Environmental Research Center more than 10 years ago, and it has been applied by Shaw Environmental, Inc. at four full-scale sites (5 systems) (Fig. 11.3).

For electrode-based proton reduction to be effective, distribution of the produced cathodic H_2 still requires the flow of groundwater for the transport to the degradative bacteria that utilize it, and this groundwater flow is limited in low permeability aquifers. Recent studies, however, have demonstrated that other H_2-producing processes may occur within clay-containing soils during the passage of an electrical current. Because these processes occur within the low permeable soils and between the electrodes, rather than on the electrode surface, the electron donor is potentially produced near the degradative bacteria, thereby reducing the amount of groundwater

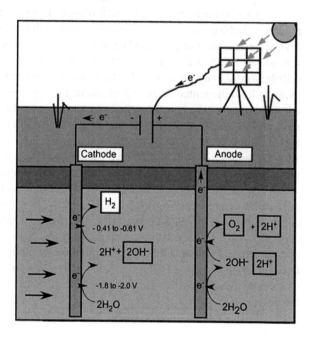

Fig. 11.2 Schematic of the proton reduction process. An electrical current is passed through the contaminated aquifer and protons (H^+) are reduced at the cathode to H_2 at a relatively low potential. The polarity of the electrodes can be reversed periodically to improve the distribution of H_2 and to control pH fluctuations. Because of the low voltage and current required, the system can be powered by a photovoltaic system

Fig. 11.3 Photo of a 215 m proton reduction system currently operating at an Oklahoma site (USA). The system functions as a permeable reactive barrier to eliminate off-site migration of chlorinated solvents, and it has been in operation successfully for >3 year

transport required for distribution of the H_2. Furthermore, these reductive clay surfaces reportedly have the ability to directly reduce chlorinated solvents in addition to protons [4]. Thus, clay-associated proton reduction has the potential to treat contaminants in low permeable soils where other technologies are ineffective.

In an early reported study of electrochemically induced reactions in soils [11] demonstrated that applying a low direct current to soils resulted in the production of "micro conductors" that acted as diluted electric chemical solid bed reactors. These micro conductors could reduce mobile oxidized chemicals in water, and reactive materials included graphite and some iron minerals. In follow-on studies by the same group [12] they showed that electrical current could be used to reduce chlorinated hydrocarbons in soils from an industrial site, and they concluded that these "micro conductors" in the soil probably play a role in the reactions.

In a much more recent study electrical current was used to reduce chlorinated solvents in clay soils at low electrical potential [4]. These researchers described the process as involving the formation of "micro capacitors" whereby hydrated clay particles become redox reactive particles and form a reactive matrix on which these redox reactions (e.g., electrolysis) can occur. They postulate that an induced electrical field in soils is created with the soil particles acting as capacitors that discharge and recharge electricity, and that can perform electrolysis of water thereby generating H_2. In addition, they reported an up to 90% reduction in TCE concentrations in the clays over 7 days under an applied current of only 6 V/m, and suggested that the decrease was due to electrically induced reduction, and not due to electro kinetic ion migration or electrophoresis. Decreases in TCE correlated with an increase in chloride concentration, and no biological daughter products were reported. This approach, therefore, appears attractive for treating chlorinated solvents trapped in clay soils, but the process has not yet been verified by field-scale demonstrations.

We anticipate that the proton reduction technology will be applicable to a wide range of treatment scenarios. Because the technology can be operated with photovoltaic power, it should be applicable to remote areas where access is limited or a constant power source is unavailable, as well sites where treatment is expected to require long treatment times (e.g., low permeability sites, large dilute plumes or large contaminant mass). It also may be suitable for treating low pH aquifers where adjusting aquifer pH can be more challenging than degrading the chlorinated solvents themselves [13]. In this case, the cathode can be used to consume hydrogen ions causing an increase in measurable pH. Finally, once validated, proton reduction may even be able to remediate chlorinated solvents sequestered in low permeability aquifer soils by utilizing H_2 production on mineral surfaces by mechanisms like those proposed by [4].

11.5 Summary

Several bioremediation technologies have been developed and proven successful for remediating chlorinated solvent-contaminated aquifers. Although indirect stimulation of native populations of DHC-type bacteria by adding a fermentable carbon source as an electron donor is likely the most common active remedial approach, the use of bio augmentation with DHC-containing bacterial culture to supplement the native bacterial population has become increasingly important and common. Despite the widespread and successful use of these technologies, many challenges remain. Often, the most challenging sites have complex hydro geologies (e.g., low permeability soils, fractured rock, or extreme pHs), or they have very high concentrations of contaminant, high contaminant mass or complex mixtures of contaminants. Consequently, new treatment technologies are still needed to address many of the remaining remediation challenges posed by these contaminants. Once such technology that has been successfully applied at full scale but continues to be developed and improved, is proton reduction facilitated by the use of electrodes and a low voltage electrical current. Continued development and testing of this technology will improve our understanding of where the technology fits among available remediation tools and where it will be the most applicable as a remediation technology. All of the treatment approaches discussed will likely continue to be widely used and improved upon to help us meet the challenges created by chlorinated solvent contamination of aquifers and to improve the quality of groundwater.

References

1. Ellis DE, Lutz EJ, Odom JM, Buchanan RJ, Bartlett CL, Lee MD, Harkness MR, Deweerd KA (2000) Bioaugmentation for accelerated in situ anaerobic bioremediation. Environ Sci Technol 34:2254–2260

2. He J, Ritalahti KM, Yang KL, Koenigsberg SS, Loeffler FE (2003) Detoxification of vinyl chloride to ethene coupled to growth of an anaerobic bacterium. Nature 424:62–65
3. Hendrickson ER, Payne JA, Young RM, Starr MG, Perry MP, Fahnestock S, Ellis DE, Ebersole RC (2002) Molecular analysis of *Dehalococcoides* 16S ribosomal DNA from chloroethene-contaminated sites throughout North America and Europe. Appl Environ Microb 68:485–495
4. Jin S, Fallgren P (2010) Electrically induced reduction of tricholoroethene in clay. J Hazard Mater 173:200–204
5. Lendvay JM, Loffler FE, Dollhopf M, Aiello MR, Daniels G, Fathepure BZ, Gebhard M, Heine R, Helton R, Shi J, Krajmalnik-Brown R, Major CL Jr, Barcelona MJ, Petrovskis E, Hickey R, Tiedje JM, Adriaens P (2003) Bioreactive barriers: A comparison of bio augmentation and bio stimulation for chlorinated solvent remediation. Environ Sci Technol 37:1422–1431
6. Lohner ST, Tiehm A (2009) Application of electrolysis to stimulate microbial reductive PCE dechlorination and oxidative VC biodegradation. Environ Sci Technol 43:7098–7104
7. Lu X, Wislon JT, Kampbell DH (2006) Relationship between *Dehalococcoides* DNA in ground water and rates of reductive dechlorination at field scale. Water Res 40:3131–3140
8. Major DW, McMaster ML, Cox EE, Edwards EA, Dworatzek SM, Hendrickson ER, Starr MG, Payne JA, Buonamici LW (2002) Field demonstration of successful bioaugmentation to achieve dechlorination of tetrachloroethene to ethene. Environ Sci Technol 36:5106–5116
9. Maymo-Gatell E, Nijenhuis I, Zinder S (2001) Reductive dechlorination of cis-1, 2-dichloroethene and vinyl chloride by *Dehalococcoides ethenogenes*. Environ Sci Technol 35:516–521
10. Maymo-Gatell X, Chien Y, Gossett JM, Zinder SH (1997) Isolation of a bacterium that reductively dechlorinates tetrachloroethene to ethane. Science 276:1568–1571
11. Rahner D, Ludwig G, Röhrs J (2002) Electrochemically induced reactions in soils—A new approach to the in-situ remediation of contaminated soils? Part 1: The micro conductor principle. Electrochem Acta 47:1395–1403
12. Röhrs J, Ludwig G, Rahner D (2002) Electrochemically induced reactions in soils—A new approach to the in-situ remediation of contaminated soils? Part 2: Remediation experiments with a natural soil containing highly chlorinated hydrocarbons. Electrochem Acta 47:1405–1414
13. Schaefer CE, Lippincott DR, Steffan RJ (2010) Field-scale evaluation of bioaugmentation dosage for treating chlorinated ethenes. Ground Water Monit R 30:113–124
14. Schaefer CE, Towne RM, Vainberg S, McCray JE, Steffan RJ (2010) Bioaugmentation for treatment of dense non-aqueous phase liquid in fractured sandstone blocks. Environ Sci Technol 44:4958–4964
15. Schaefer SE, Condee CW, Vainberg S, Steffan RJ (2009) Bioaugmentation for chlorinated ethenes using *Dehalococcoides* sp.: Comparison between batch and column experiments. Chemosphere 75:141–148
16. Vainberg S, Condee CW, Steffan RJ (2009) Large scale production of *Dehalococcoides sp.*-containing cultures for bioaugmentation. J Indust Microbiol Biotechnol 36:1189–1197
17. Weathers LJ, Parkin GF, Alvarez PJ (1997) Utilization of cathodic hydrogen as electron donor for chloroform cometabolism by a mixed, methanogenic culture. Environ Sci Technol 37: 880–885

Chapter 12
Bioremediation of Petroleum Contaminated Water and Soils in Tunisia

Boutheina Gargouri, Sami Mnif, Fathi Aloui, Fatma Karray, Najla Mhiri, Mohamed Chamkha, and Sami Sayadi

Abstract The petrochemical industry generates series of liquid and solid wastes containing large amounts of priority pollutants during the petroleum-refining process. These residues must be treated through depuration processes. The bioremediation process, presenting countless advantages in relation to other processes employed, is an evolving method for the removal and the transformation of many environmental pollutants including those produced by the petroleum industry. In a first step, a continuously stirred tank bioreactor (CSTR) was used to optimize feasible and reliable bioprocess system for successful bioremediation of industrial effluent and to develop an efficient microbial consortium for the degradation of petroleum hydrocarbons. After an experimental period of 175 days, the process was shown to be highly efficient in decontaminating the wastewater. The performance of the bio augmented reactor was demonstrated by the reduction of COD rates up to 95%. Six microbial isolates from the CSTR were characterized and species identification was confirmed by sequencing the 16 S rRNA genes. Besides, the treated wastewater could be considered as non toxic according to the micro-toxicity test. In a second step, bioremediation of a refinery soil containing hydrocarbons climate was investigated. The objective of this study was to assess the ability of bioremediation technique in the presence of the acclimatized consortium to reduce the total petroleum hydrocarbon (TPH) content in the contaminated soil. Results clearly demonstrated that an enhanced bioremediation was carried when the acclimatized bacterial consortium was added to the hydrocarbons contaminated soil. The proposed bioremediation technology has proved significantly higher hydrocarbons removal efficiencies. TPH analysis showed that 50% of the hydrocarbons were

B. Gargouri • S. Mnif • F. Aloui • F. Karray • N. Mhiri • M. Chamkha • S. Sayadi (✉)
Laboratoire des Bioprocédés Environnementaux, Pôle d'Excellence Régional AUF-LBPE,
Centre de Biotechnologie de Sfax, Université de Sfax, BP 1117, 3018 Sfax, Tunisia
e-mail: sami.sayadi@cbs.rnrt.tn

F.F. Quercia and D. Vidojevic (eds.), *Clean Soil and Safe Water*, NATO Science for Peace
and Security Series C: Environmental Security, DOI 10.1007/978-94-007-2240-8_12,
© Springer Science+Business Media B.V. 2012

eliminated during the first 15 days of bio remediation. TPH removal reached 96% at the end of the treatment. Further, GC/MS profile has proved that the acclimatized bacterial consortium could effectively remove the medium- and long-chain alkanes in the contaminated soil such as the alkanes were undetectable after a 30-day of incubation period. In a third step, a *Halomonas* sp. strain C2SS100 had been isolated and characterized from Sercina petroleum reservoir. The strain had shown potential hydrocarbon degradation under halophilic condition (100 g 1^{-1} NaCl). During growth on n-Hexadecane (C_{16}), C2SS100 produced biosurfactant that could solubilise phenanthrene, a three-ring aromatic hydrocarbon. The halophilic character of this bacterium could add further advantages for its use in marine and saline environments-oil bioremediation.

12.1 Effects of Wastewater Organic Loading Rates (OLR) and the Hydrolytic Retention Time (HRT) Variations on the Reactor Performances

The CSTR bioreactor was operated under six HRTs (Fig. 12.1). The purpose of these stages was to evaluate the functional performance of the bioreactor in terms of hydrocarbon degradation. The feeding flow was progressively increased from 8, 12, 15 to 18 L D^{-1}, corresponding to a HRT decrease from 6 days to 16 h. The reactor performance was investigated by COD, BOD_5, pH and biomass measurement. At HRT of 6 days and 4 days, the reactor maintained an average of $COD_{effluent}$ and $BOD_{5effluent}$ concentrations which were 900 mg L^{-1} and 240 mg L^{-1}, respectively, corresponding to an ORL of 0.36 g COD L^{-1} D^{-1} Decreasing the HRT from 1 day to 21 h and then after to 16 h, affected significantly the reactor performances. Indeed, after continuous treatment in the CSTR, the $COD_{effluent}$ and $BOD_{5effluent}$ average removals were high reaching 96% and 93%, respectively (Fig. 12.1a, b). For the first three OLR, the pH of the CSTR effluent increased from 7.2 to 8. The pH values are, therefore, within the acceptable range proper reactor operations as were previously described [6].

As shown in Fig. 12.1c, the initial biomass of 1.2 g L^{-1} remained stable until day 65. During the last two OLR, the biomass concentration increased slightly and reached 3.5 g VSS L^{-1}. Then, it increased with time and reached 7.9 g L^{-1} in terms of VSS, at the end of the treatment.

12.2 LUMIStox Test

V. fischeri bioluminescence test was used to evaluate the toxicity of treated wastewater before their discharge into the environment as was indicated by [7]. Several wastewater samples were analysed for their inhibitory potential of the well-known bacterial strain *V. fischeri*. Figure 12.2 shows that the percentage of

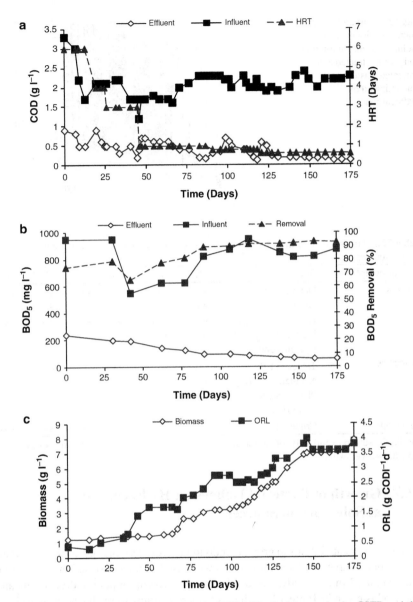

Fig. 12.1 Variations in the filtered COD (**a**), the BOD$_5$ (**b**), the biomass in the CSTR and the organic load evolutions (**c**) during biodegradation wastewater treatment in the CSTR

inhibition of this strain varied from 60% to 68% for untreated wastewater samples and bioluminescence ranged from 37% to 0.4% at the end of treatment process, which represents 99% removal of the initial toxicity. This decrease was due to the efficient conversing of the toxic raw-petroleum during biodegradation.

Fig. 12.2 Variations of the
micro-toxicity expressed as
bioluminescence inhibition
($\%I_B$) of the treated and
untreated wastewater
contaminated by
hydrocarbons

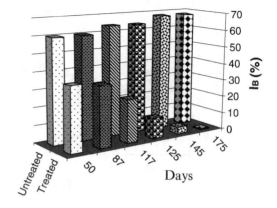

Table 12.1 Growth and ability of isolated strains to use different hydrocarbon compounds as carbon sources

Organic compounds	Strain	HC2	HC5	HC6	HC7	HC8	HC9
Alkanes	Octane (C_8)	+	++	+	++	+	++
	Decane (C_{10})	++	++	++	++	++	++
	Hexadecane (C_{16})	++	+++	++	++	++	+++
	Eicosane (C_{20})	+	+	++	+	++	++
	Cyclohexane	+	+	+	+	+	+
Aromatics	Toluene (1 ring)	-	+	-	-	-	-
	Naphtalene (2 rings)	-	-	-	-	-	-
	Phenanthrene (3 rings)	-	-	-	-	-	+
Heavy oil	Crude oil	++	+++	++	++	+++	+++

For liquid substrate turbidity of cultures: + = low growth; ++ = medium growth; +++ = strong growth; - = no growth

12.3   Growth of Bacterial Isolates on Hydrocarbons as Sole Carbon Sources

Based on their capabilities to grow on individual hydrocarbons and crude oil as their sole carbon source, six bacteria isolates were used in the construction of bacterial consortia used in this study. Result of bacterial growth on pure hydrocarbons and heavy oil is given in Table 12.1. The six bacterial strains were able to grow well on aliphatic hydrocarbons such as hexadecane (C_{16}), and eicosane (C_{20}) as well as on heavy oil (crude oil). Also, they grew poorly on octane (C_8), and cyclohexane. However, these bacterial strains did not appear to utilize aromatic hydrocarbons as carbon source, except strain HC9 which grew poorly on phenanthrene and strain HC5 which utilized toluene poorly. Among the six bacterial strains tested in hydrocarbon utilizations, strain HC9 was the most powerful hydrocarbon degrader. Therefore, this strain will be selected for future studies, especially in investigations on the environmental factors influencing the bioremediation of the oil spill.

Table 12.2 Identification of bacterial isolates on the basis of biochemical tests and 16 S rDNA sequencing O/F: oxidative/fermentative

Strains	Relevant characteristics[a]	Identification[b]	% Similarity
HC2	Gram (−), Catalase (+), motile, O⁺/F⁺, Rhodamine (+)	*Aeromonas punctata*	99
HC5	Gram (+), Catalase (+), unmotile, O⁻/F⁻, Rhodamine (−)	*Bacillus cereus*	99
HC6	Gram (−), Catalase (+), motile, O⁻/F⁻, Rhodamine (−)	*Ochrobactrum intermedium*	99
HC7	Gram (−), Catalase (+), motile, O⁺/F⁻, Rhodamine (−)	*Achromobacter spp.*	99
HC8	Gram (−), Catalase (+), motile, O⁺/F⁻, Rhodamine (−)	*Stenotrophomonas maltophilia*	99
HC9	Gram (+), Catalase (+), unmotile, O⁺/F⁻, Rhodamine (+)	*Rodococcus sp.*	99

[a]As determined by biochemical tests (API)
[b]As confirmed by 16 S rDNA sequencing

12.4   Identification of Petroleum-Degrading Bacteria Isolated from Enrichment Process

Six aerobic bacteria were isolated from petroleum hydrocarbon-contaminated water using industrial wastewater as the sole carbon source. The isolated bacterial cultures were characterized by their morphological and biochemical properties (Table 12.2). According to the data obtained using light and electron microscopy, the isolated bacteria had the shapes of rods or cocci, were spore-forming or non-spore-forming, occurred as single cells or were integrated in chains, and were immotile or motile with flagella. Two strains are Gram-positive and four strains are Gram-negative.

Specie identifications were performed using biochemical test (API) and confirmed by sequencing the 16 S rRNA genes of the six isolates (Table 12.2). The partial 16 S rRNA genes sequencing of HC2, 5, 6, 7, 8 and 9 strains were determined (data not shown).

The sequence comparison demonstrated the affiliation of most strains (HC2, 6, 7 and 8) to *Proteobacteria* phylum. HC2, 5, 6, 8 and 9 strains were closely related to *Aeromonas punctata* (*Aeromonas caviae*), Bacillus cereus, *Ochrobactrum intermedium, Stenotrophomonas maltophilia* and *Rhodococcus* sp. respectively. HC7 isolate was affiliated to genus *Achromobacter* with a similarity of 99%. Bacteria of the genera *Aeromonas, Achromobacter, Rhodococcus, Stenotrophomonas, Ochrobactrum* and Bacillus, isolated in this study, belonged to bacteria, playing an important role in the degradation of petroleum hydrocarbons in the environment, as previously reported [8].

The most successful process for the removal or the elimination of hydrocarbon from the environment is the microbial transformation and biodegradation. The rate of biodegradation of petroleum hydrocarbon in the environment depends on different factors such as pH, temperature, oxygen, microbial population and chemical structure of the compounds [3, 9]. Many studies were focused on the isolation and characterisation of microorganisms degrading hydrocarbon components. Numerous microorganisms, namely bacteria, yeast and fungi have been reported as good degraders of hydrocarbons [11]. Many of these microorganisms were applied in bioremediation processes to reduce the concentration and the toxicity of various pollutants, including petroleum products.

12.5   Biodegradation of TPH in Soil Microcosms

Figure 12.3 shows the level of TPH biodegradation in soil microcosms (sand, gravel and finer particles, pH 7.8 with 8.5% organic carbon and 0.16% total nitrogen) over 30 days. Initial TPH concentration of soil was 63.4 mg g^{-1} as determined by gravimetric analysis.

The Comparison of the chromatographic profiles in non-inoculated control and in the bio augmented soil sample with acclimatized consortium showed a high disappearance of the different hydrocarbons in representative bio augmented one.

The kinetics observed in the decline of TPH during the bioremediation process was analysed. Control soil samples, without microbial consortium, showed a delay in the reduction of TPH concentration, which gradually decreased to 40.5 mg g^{-1} (35%) of the initial concentration of hydrocarbons was removed after 30 days. However, microcosms that involved acclimatized microbial addition led to a large decrease in TPH by over 92%. The significant TPH reduction observed in bio augmented samples could be related to the soil proprieties and soil hydrocarbon degradation potential [4].

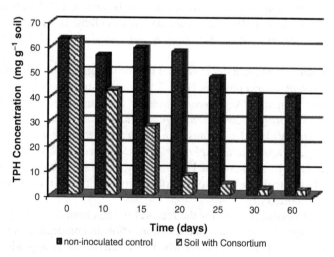

Fig. 12.3 TPH degradation levels in soil with acclimatized consortium and in the control plots

12.6   Change in GC/MS Profiles during Bioremediation of Hydrocarbon-Contaminated Soil

Table 12.3 presents the compounds detected by GC/MS analysis of extracts of the various soil samples excavated from an industrial hydrocarbon-contaminated area. The data reflect the fact that the soil used in this investigation was sampled from the dumping area of an industrial Tunisian refinery where the initial pollutants were of very diverse composition, i.e. a mixture of crude oil, mazout, diesel, middle distillates, heavy distillates, kerosene, etc. [2]. The untreated soil sample contained a large variety of straight-chain hydrocarbons and their methyl derivatives.

The various classes of chemical detected by GC/MS analysis were in agreement with the literature data for soils contaminated with crude and oil derivatives [1]. Table 12.3 shows that abundance of most substances in the hydrocarbons contaminated soil greatly decreased during bioremediation process. This means that a significant improvement of the soil quality was achieved, confirming the ability of the acclimatized consortium to decompose the organic compounds of the contaminated soil.

Table 12.3 Compounds detected in soil samples in the course of bioremediation

| Hydrocarbons | T_R | M^+ abundance of hydrocarbons | | | |
Days of the experiment		0	15	20	30
Dodecene	6.3	168	++	+	−	−
Dodecane (C_{12})	6.4	170	++	+	−	−
Tetradecene	7.8	196	++	+	−	−
Teradecane(C_{14})	7.8	198	++	+	−	−
Hexadecene	9.0	224	++	+	−	−
Hexadecane (C_{16})	9.1	226	++	+	−	−
Heptadecen	10.2	238	++	+	−	−
Heptadecane (C_{17})	10.2	240	++	+	−	−
Cyclohexadecane	11.0	251	−	+	+	−
Octadecene	11.0	252	++	+	+	−
Octadecane (C_{18})	11.1	254	++	+	+	−
Nonadecane (C_{19})	11.3	268	−	+	−	−
Eicosane (C_{20})	11.8	282	++	+	+	−
Hineicosane (C_{21})	12.0	296	−	+	+	−
Docosane (C_{22})	12.4	310	++	+	+	−
Tricosane (C_{23})	12.0	324	−	+	−	−
Cycloteracosane	12.7	336	−	+	−	−
Tetracosane (C_{24})	12.4	338	++	+	+	−
Pentacosane (C_{25})	12.7	352	−	+	−	−
Heptacosane(C_{27})	13.5	380	++	+	−	−
Octacosane (C_{28})	13.9	394	−	+	−	−
Triacontane (C_{30})	15.0	422	++	+	−	−

(++) very good abundance, (+) good abundance, (−) no band observed

In conclusion, our results showed that acclimatized microbial consortium was able to degrade an important fraction of petroleum compounds and that bio augmentation could be a valuable alternative tool to improve bioremediation. In addition, the laboratory experiments were effectively extrapolated to the field scale in a petroleum industry in Tunisia. The treatment provided satisfactory results and will present a feasible technology for the treatment of hydrocarbon-rich wastewater and contaminated soil from petrochemical industries and petroleum refineries.

12.7   A Halophilic Bacterium *Halomonas sp.* C2SS100, Growing on Crude Oil and *n*-Hexadecane, Produced a New Salt Tolerant Biosurfactant

Production water collected from "Sercina" petroleum reservoir, located near the Kerkennah Island in Tunisia, was used for the screening of halotolerant or halophilic bacteria able to degrade crude oil. Bacterial strain C2SS100 was isolated after enrichment on crude oil, in the presence of 100 g L^{-1} NaCl and at 37°C. This strain was aerobic, Gram-negative, rod-shaped, motile, oxidase + and catalase +. Phenotypic characters and phylogenetic analysis based on the 16 S rRNA gene of the isolate C2SS100 showed that it was related to members of the *Halomonas* genus. The degradation of several compounds present in crude oil was confirmed by GC-MS analysis. The use of refined petroleum products such as diesel fuel and lubricating oil as sole carbon source, under the same conditions of temperature and salinity, showed that significant amounts of these heterogenic compounds could be degraded. Strain C2SS100 was able to degrade *n*-hexadecane (C$_{16}$). During growth on hexadecane, cells surface hydrophobicity and emulsifying activity increased indicating the production of biosurfactant by strain C2SS100 [10]. Phenanthrene solubility in water was enhanced by biosurfactant addition. Our results suggest that the C2SS100 biosurfactant has interesting properties for its application in bioremediation of saline environment polluted with hydrocarbon compounds.

12.7.1   Crude Oil, Diesel Fuel and Lubricating Oil Biodegradation

C2SS100 was able to degrade crude oil (1% v/v) in basal liquid and solid media, in the presence of 100 g L^{-1} NaCl. The growth of the strain on crude oil was followed by measuring the OD$_{600\ nm}$ at different culture's times. GC-MS analysis showed that strain C2SS100 was active on the total aliphatic hydrocarbons present in crude oil (C$_{11}$–C$_{22}$) after 30 days of incubation (Fig. 12.4). This result was confirmed by diminution or total disappearance of the correspondent peak of each compound. The strain C2SS100 degrades higher than 90% of crude oil present in the culture after four weeks incubation (Fig. 12.4).

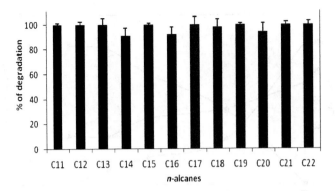

Fig. 12.4 Degradation of *n*-alkanes by strain C2SS100 after 30 days of exposure at 150 rpm, and in the presence of 100 gL^{-1} 37°C. All results of biodegradation were calculated as the difference between the controls and the inoculated flasks containing 50 ml of basal medium and crude oil (1%, v/v). C_{11} to C_{22} indicate n-alkanes with the number of carbon atoms from 11 to 22

Fig. 12.5 Growth of strain C2SS100 on (♦) crude oil (1% v/v), (■) diesel fuel (1% v/v) and (▲) lubricating oil (1% v/v) determined by bacterial cell counts

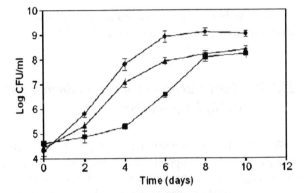

Addition of tween 80 showed a significant acceleration of crude oil degradation. In fact, the growth curve of C2SS100 showed that stationary phase was reached after about 3 days of incubation only in the presence of tween 80, whereas in the absence of this surfactant, one week was required. The crude oil toxicity for growth of the strain C2SS100 was also studied by measuring the strain growth at a crude oil concentration range between 0 to 40% (v/v). Strain C2SS100 can grow even in presence of 20% (v/v) crude oil. The optimum growth was observed in the presence of 2% (v/v) of crude oil. At 40% (v/v), crude oil became toxic and inhibited the growth of strain C2SS100 (data not shown).

Crude oil was the best substrate to support bacterial growth, and strain C2SS100 reached the stationary phase within 6 days. In contrast, the growth of C2SS100 in the presence of lubricating oil and diesel fuel showed extended lag times, and approximately 7 and 8 days, respectively, were required to reach stationary phase (Fig. 12.5). These results were confirmed also by enumeration of viable cells, indicating positive growth with the three substrates (Fig. 12.5).

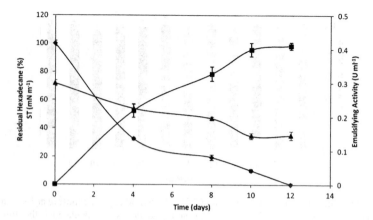

Fig. 12.6 Evolution of emulsifying activity and surface tension during degradation of hexadecane by strain C2SS100. (▲): Surface tension, (■): Emulsifying activity and (♦): Percentage of residual hexadecane

These findings suggest that strain C2SS100 had substrate preference as follows: crude oil > lubricating oil > diesel fuel (Fig. 12.5).

12.7.2 Biosurfactant Production during N-Hexadecane (C16) Degradation

C2SS100 was able to grow aerobically at 37°C, in the basal medium containing 100 g L^{-1} NaCl and supplemented with 0.5% (v/v) hexadecane as the sole carbon and energy source. The concentration of hexadecane in the culture decreased until substrate utilization was complete. Hexadecane was completely degraded after 12 days of incubation (Fig. 12.6).

After 6 days of incubation, GC-MS profile showed the appearance of a new peak corresponding to hexadecanoic acid. The new peak of this intermediate was maintained even at 8 days of incubation but disappeared after 10 days of incubation. During growth on hexadecane, emulsifying activity was increased and surface tension was decreased to 34.7 mN m^{-1} (Fig. 12.6). This result suggests that the strain has the capacity to produce biosurfactant that enhanced hydrocarbon degradation.

12.7.3 C2SS100 Biosurfactant Application in Phenanthrene Solubilisation

Phenanthrene, a three-ring PAH (Polycyclic Aromatic Hydrocarbon) with water solubility as low as 6.6×10^{-6} mol L^{-1} [13], was selected here as representative of PAHs to determine the solubilisation caused by C2SS100 biosurfactant solution.

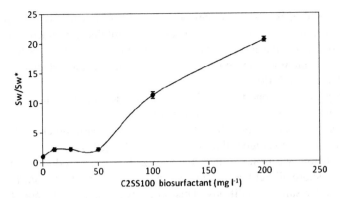

Fig. 12.7 Solubilisation of phenanthrene by 1E biosurfactant in pure water supplemented with 100 g L^{-1} NaCl

6 mg phenanthrene (purity 98%, Aldrich-Germany) was added into a series of 10 ml biosurfactant solutions at different concentrations (prepared in distilled water with 100 g L^{-1} NaCl).

This concentration of phenanthrene is much higher than the solubility of phenanthrene in saline water (100 g L^{-1}). The concentrations of biosurfactant solution were adjusted to 12.5, 25, 50, 100 and 200 mg L^{-1} individually. Each concentration was performed in triplicate and saline water was served as control. The solubilisation experiments were performed in 20 ml centrifuge tubes. These tubes, which contained the mixture of phenanthrene and water, were agitated in a vertical position at 30°C with shaking (200 rpm) in the dark for 24 h, followed by centrifugation at 12,500 rpm. at 4°C for 30 min. A 5-ml aliquot of supernatant was collected; the pH was adjusted to 2.0 using 6 M HCl. Then an equal volume of dichloromethane was added in a separation funnel to extract phenanthrene dissolved in biosurfactant. This process was repeated twice. Afterward, the organic portions were combined, dried with anhydrous sodium sulphate and concentrated to 1 ml. The concentrated liquid was filtered through a 0.20 mm micro-pore filter before being analyzed by UV–vis spectrophotometer at 254 nm (Shumazu-1,100, Japan) according to [5].

PAHs, one of the persistent organic pollutants found in oil containing wastewater, are also a kind of hydrophobic organic compounds (HOCs) [12]. As well, the water solubility of PAHs decreases with the increasing number of rings in molecular structure. This property induces the low bioavailability of these organic compounds that is an important factor in the biodegradation of these compounds. The water solubility of some HOCs can be improved by surfactant or biosurfactant owing to its amphipathic structure. The degree of solubilisation caused by biosurfactant can be expressed through the ratio of S_w^* (the apparent solubility of a certain solute in surfactant solution) to S_w (the solubility of a certain solute in pure water). As shown in Fig. 12.7, biosurfactant has a highly obvious effect on solubilisation of phenanthrene. The solubility of phenanthrene in biosurfactant solution was about 20 times higher than the control solubility when the concentration of biosurfactant was at its highest point. Our results also revealed that biosurfactant had a certain effect on the solubilisation of phenanthrene either below or above its CMC (the CMC of this biosurfactant was 150 mg L^{-1}). Nevertheless, the solubilisation function was much more obvious when the biosurfactant concentration was above its CMC. This is because surfactant molecules exist as monomers when the biosurfactant concentration is below its CMC [5].

Since this kind of monomer has little partition effect on the solute, solubility of phenanthrene was very small or slightly increased.

However, the apparent solubility of phenanthrene increases quickly when biosurfactant concentration is higher than 150 mg L^{-1} (200 mg L^{-1}).

In conclusion, we isolated and characterized, from a production water of a Tunisian off-shore oil field, a *Halomonas* sp. strain C2SS100 after enrichment on crude oil under saline conditions. This strain was capable of degrading some aliphatic hydrocarbons present in crude oil efficiently. During growth on hexadecane (C$_{16}$), strain C2SS100 produced biosurfactant with potential application on phenanthrene solubilisation under halophilic conditions (100 g L^{-1}). The halophilic character of this bacterium could add further advantages for its use in marine and saline environments-oil bioremediation.

References

1. Aboul-Kassim TAT, Simoneit BRT (2001) Organic pollutants in aqueous-solid phase environments: Types, analyses and characterizations. The handbook of environmental chemistry. In: Pollutant-solid phase interactions: mechanism, chemistry and modelling, 5 part Eth edn. Springer, Berlin/Heidelberg
2. Anonymous (1997) Innovations in groundwater and soil cleanup: from concept to commercialisation, CIRT-WSTB-BRWM-CGER-NRC. National Academy Press, Washington, DC
3. Atlas RM (1981) Microbial degradation of petroleum hydrocarbons: environmental perspective. Microb Rev 45:180–209
4. Bento FM, Camargo FAO, Okeke BC, Frankenberger WT (2005) Comparative bioremediation of soils contaminated with diesel oil by natural attenuation, biostimulation and bioaugmentation. Bio resource Technol 96:1049–1055
5. Cuny P, Faucet J, Acquaviva M, Bertrand JC, Gilewicz M (1999) Enhanced biodegradation of phenanthrene by a marine bacterium in presence of a synthetic surfactant. Lett Appl Microbiol 29:242–245
6. Gargouri B, Karray F, Mhiri N, Aloui F, Sayadi S (2011) Application of a continuously stirred tank bioreactor (CSTR) for bioremediation of hydrocarbon-rich industrial wastewater effluents. J Hazard Mater 189:427–434
7. Hao OJ, Chien-Jen S, Chen-Fang L, Fu-Tien J, Zen-Chyuan C (1996) Use of microtox tests for screening industrial wastewater toxicity. Water Sci Technol 34:43–50
8. Juhas AL, Stanley GA, Britz ML (2000) Microbial degradation and detoxification of high molecular weight polycyclic aromatic hydrocarbons by Stenotrophomonas maltophilia strain VUN 10,003. Lett Appl Microbiol 30:396–401
9. Leahy JG, Colwell RR (1990) Microbial degradation of hydrocarbons in the environment. Microb Rev 54:305–315
10. Mnif S, Chamkha M, Sayadi S (2009) Isolation and characterization of Halomonas sp. strain C2SS100, a hydrocarbon-degrading bacterium under hyper saline conditions. J Appl Microbiol 107:785–794

11. Samanta KS, Singh OV, Jain RK (2002) Polycyclic aromatic hydrocarbons: Environmental pollution and bioremediation. Trends Biotechnol 20:243–248
12. Shin KH, Kim KW, Ahn Y (2006) Use of bio surfactant to remediate phenanthrene contaminated soil by the combined solubilisation-biodegradation process. J Hazard Mater 137:1831–1837
13. Zhu LZ, Feng SL (2002) Water solubility enhancement of polycyclic aromatic hydrocarbons by mixed surfactant solutions. Acta Sci Circle 22:774–8

Chapter 13
Remediation of Metal Ion-Contaminated Groundwater and Soil Using Nanocarbon-Polymer Composition

Rashid A. Khaydarov, Renat R. Khaydarov, Olga Gapurova, and Radek Malish

Abstract The presence of different organic and heavy metal contaminants in groundwater and soil has a large environmental, public health and economic impact. The paper deals with a novel method of groundwater and soil remediation using nanocarbon-polymer composition (NCPC). The process of NCPC synthesis and its chemical characteristics have been described. Nano-carbon colloids (NCC) and polyethylenimine (PEI) are used to synthesis of NCPC. Metal ions interact with NCPC via ion exchange and complexation mechanism. The ability to remove metal ions from water against pH, ratio of NCC and PEI in NCPC, speed of coagulation of NCPC and size of NCC has been investigated. NCPC has a high bonding capacity of 4.0–5.7 mmol/g at pH 6 for most divalent metal ions. The percent of sorption of Zn (II), Cd (II), Cu (II), Hg (II), Ni (II) and Cr (VI) ions is higher than 99%, and distribution coefficients are 10^1–10^3. The lifetime of NCPC before coagulation in the treated water and soil is 1 s to 1,000 min and depends on the ratio of polymeric molecules and carbon nanoparticle concentrations. Accordingly, the depth of penetration of NCPC in a soil or depth of remediation of soil can change from 1 to 100 cm, and distance of moving the NCPC with groundwater or remediation zone of ground can change from 1 to 100 m. Thus NCPC can be used for effective removal of metal ions from contaminated water and remediation of soil. The results of field tests of the method have been also described.

R.A. Khaydarov (✉) • R.R. Khaydarov • O. Gapurova
Institute of Nuclear Physics, Ulugbek, 100214 Tashkent, Uzbekistan
e-mail: renat2@gmail.com

R. Malish
Institute of Tropics and Subtropics, Prague, Czech Republic

F.F. Quercia and D. Vidojevic (eds.), *Clean Soil and Safe Water*, NATO Science for Peace and Security Series C: Environmental Security, DOI 10.1007/978-94-007-2240-8_13,
© Springer Science+Business Media B.V. 2012

13.1   Introduction

An extensive use of metals and chemicals in process industry has resulted in generation of a large amount of effluent containing high levels of toxic heavy metals. Furthermore mining and mineral processing procedures are known to generate toxic liquid waste. The presence of different organic and heavy metal contaminants in groundwater and soil has a large environmental, public health and economic impact. Most of traditional methods such as solvent extraction, activated carbon adsorption, ion exchange, biological degradation and precipitation techniques are rather efficient, but often are costly and/or time-consuming.

Authors of [24] were the first to introduce the method of separation by coupling ultra filtration and complexation of metallic species with industrial water soluble polymers. This approach implies that polymeric molecules operate as a sorbent and a carrier of metal ions. In the method metal ions are primarily bound to water-soluble polymers. The unbound ions pass through the membrane, whereas the polymers and their complexes are retained. Within this approach various weakly basic, water-soluble polymers including chitosan, polyethylenimine (PEI), poly (diallyldimethylammonium chloride), sodium polystyrene sulfonate (PSS) etc. have been used for removal metal ions from water [1, 5–7, 21, 25, 26]. Main advantages of the method lie in relatively low energy requirements of the ultra filtration process and high removal efficiency due to effective binding of metal ions with polymers [6]. On the other hand, there is a difficulty of removal of the carrier of metal ions from concentrated solution and the separation of polymeric molecules and metal ions.

Application of carbon nanoparticles and nanotubes for removal of metal ions (Cd^{2+}, Cu^{2+}, Ni^{2+}, Pb^{2+}, Zn^{2+}, etc.) from water have been described recently [4, 23]. Within this approach nanoparticles or nanotubes operate as a sorbent and a carrier of metal ions. Electrolytically generated nanocarbon colloids (NCC) have functional groups such as carbonyl, hydroxyl and carboxyl groups forming on the surface of carbon nanoparticles [3, 9, 10, 22]. The sorption capacities of these particles are high, up to 7 mmol/g [8], and is purely comparable with that of cation-exchange resins. On the contrary, the sorption capacities of raw nanotubes for metal ions are very low, because walls of carbon nanotubes are not reactive. But their fullerene-like tips are known to be more reactive, and the sorption capacity significantly increases after oxidation by HNO_3, NaOCl and $KMnO_4$ solutions due to the generation of –COOH, –OH, or –CO groups [2, 11–20]. The sorption capacities of carbon nanotubes are not greater than 1 mmol/g [23] that is 2–5 times less than those of cation-exchange resin. The main disadvantage of this method is a complexity of separation of metal ions carriers (i.e. nanoparticles with metal ions) from water. In order to improve the separation of carriers of metal ions from treated water, the metal ions can be bound to polymeric molecules and carbon nanoparticles forming nanocarbon-conjugated polymer nanocomposites (NCPC) in water that is able to precipitate rapidly.

The idea of the technique described in the paper is to add a water-soluble polymeric molecules and carbon colloids to water in such a way as to bind metal ions and simultaneously form NCPC. That leads to significant increase of the size of NCPC species with follow-up formation of precipitates. This sediment can be easily removed from water by filtration or centrifugation with follow-up procedure of extraction of the metals.

13.2 Materials and Equipment

Copper sulphate pentahydrate ($CuSO_4·5H_2O$), cobalt nitrate hexahydrate (Co $(NO_3)_2·6H_2O$), nickel nitrate hexahydrate (Ni $(NO_3)_2·6H_2O$), cadmium nitrate tetra hydrate (Cd $(NO_3)_2·4H_2O$), zinc nitrate hexahydrate (Zn $(NO_3)_2·6H_2O$), potassium chromate (K_2CrO_4), mercury chloride ($HgCl_2$), Poly (ethylenimine) (molar weight of 10,000 and 200,000–350,000) from Sigma Aldrich, UK were used in the as received condition.

NCC was prepared by the electrochemical method that we described earlier [8]. The process was based on the use of two-electrode device in which an anode and a cathode are made from high-density isotropic graphite's OEG4 (Russia) (65 mm × 30 mm × 15 mm) to be transformed into carbon colloidal particles. The anode and the cathode were immersed in a plastic electrolytic cell (120 mm × 140 mm × 105 mm) filled with distilled or deionised water as the electrolyte. The distance between the electrodes was varied from 10 to 120 mm in the current density range 0.1–3 mA/cm² at a constant voltage of 60 V. The electrolytic cell was installed on the magnetic stirrer in order to provide water flowing between the electrodes. That allows saturating the electrolyte with carbon colloids and discharging the gas generated on the surface of electrodes due to electrolysis of water. The process of device operation involved two repeatable consecutive steps: (1) the electrolysis during 10 min, (2) the electrolyte stirring during 60 s. The process was executed automatically, using twin timer ST-T (South Korea).

The NCC preparation process was executed by two stages: anode activation and the carbon nanoparticles generation. At the first stage the electrolyte has low conductivity, value of electric current density is small, about 0.1–0.2 mA/cm² and the oxidation reaction is slow. Duration of this stage is about 50 h and depends on the quality (density) of graphite. At this stage a voltage between electrodes is high, about 60–100 V. As the reaction proceeds, the conductivity of the electrolyte is abruptly increased, the current density increases up to 10 mA/cm² and higher and the oxidization reaction sets in. As a result, the carbon electrode is finely split with follow-up covering by the carboxyl group. At the second stage the electric current density between electrodes is about 3–4 mA/cm². The NCC was stable during at least 150 days. The ion exchange capacity of NCC was about 7 m mol/g for H^+ - ions.

The size and the shape of nanoparticles were determined with transmission electron microscopy (TEM) (LEO-912-OMEGA, Carl Zeiss, and Germany). The size values were averaged over more than 200 nanoparticles from different TEM micrographs of the same sample. Conductivity and pH of solutions were measured with WTW bench Multi-parameter Multi-Lab 540.

Table 13.1 Radio nuclides used as labels ($T_{1/2}$ – half-life of the radio nuclides, $E\gamma$ – energy of the γ-peak)

Elements	Radio nuclides	$T_{1/2}$	E_γ, MeV
Cr(VI)	^{51}Cr	27.73 days	0.320
Ni(II)	^{65}Ni	2.5 h	1.480
Cu(II)	^{64}Cu	12.7 h	0.511
Zn(II)	^{65}Zn	244.1 days	1.115
Cd(II)	^{115}Cd	53.5 h	0.336
Hg(II)	^{203}Hg	46.6 days	0.279

The concentration of PEI in solutions was determined by neutron activation analysis. Water samples were irradiated by the Fast Neutron Generator of the Institute of Nuclear Physics (Tashkent, Uzbekistan). NaJ (Tl) 63 × 63 mm detector and a 6,144-channel multichannel analyzer were used for recording gamma-ray quanta. The area under γ-peak of radionuclide 13 N (half-life $T_{1/2}=9.96$ min, energy of the γ-peak $E_\gamma=0.511$ MeV) was measured to determine the concentrations of nitrogen. Based on the obtained values the concentration of PEI in water was calculated.

Radio nuclides used as the label of ions during the study of water purification process are given in Table 13.1. The radio nuclides were prepared by irradiating salts of ions in the nuclear reactor of the Institute of Nuclear Physics (Tashkent, Uzbekistan).

Ge (Li) detector with a resolution of about 1.9 keV at 1.33 MeV and a 4,096-channel analyzer were used for detection gamma-ray quanta. Areas under γ-peaks of radio nuclides were measured to calculate the amount of ions.

The bonding capacity Q, mmol/g, was calculated as follows:

$$Q = \frac{(A_0 - A_e) * B}{(A_0 - A_B) * W} \qquad (13.1)$$

Where B is amount of metal ions, mmol; W is a weight of absorber, g; A_0 is a count rate value of the initial solution, A_e is a count rate of the solution at equilibrium, A_B is a background count. The distribution coefficient K_d and the removal ratio P were calculated by Eqs. 13.2 and 13.3:

$$K_d = \frac{A_0 - A_e}{A_e - A_B} * \frac{V}{W} \qquad (13.2)$$

$$P = \frac{A_0 - A_e}{A_0 - A_B} * 100 \qquad (13.3)$$

Where V is a total volume of the solution, ml

For studies of NCPC bonding capacity, distribution coefficient and the removal ratio solutions containing NCPC were passed through a syringe filters with a pore size of 20 nm (Dubna, Russia).

13.3 Results and Discussions

The process of metal ions removal from water comprises the following stages: (i) capture of metal ions by PEI and NCC and simultaneous formation of NCPC, (ii) sedimentation of NCPC, (iii) removal of sediments containing NCPC with metal ions by filtration [9].

13.3.1 Formation of NCPC

Two methods of NCPC formation have been studied: (a) the NCPC formation by adding PEI solution with the concentration from 60 mg/l to 320 mg/l into the NCC solution in the concentration range 2–100 mg/l; (b) the NCPC formation by adding a solution of NCC with the concentration from 100 mg/l to 10,000 mg/l into the PEI solution in the concentration range 50–320 mg/l.

The typical TEM image for NCC obtained is shown in Fig. 13.1. pH and concentration ranges of NCC were 2.8–3.1 and 150–400 ppm, respectively, and depends on the duration of the process.

Equilibrium reaction for water solution of NCC can be written as:

$$Carb - kCOOH \leftrightarrow Carb - kCOO^- + kH^+ \qquad (13.4)$$

Where $0 \leq k \leq k_m$ (k_m is the number of carboxyl groups on the surface of carbon nanoparticle).

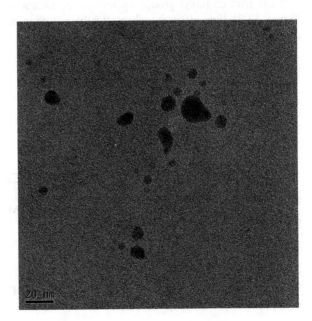

Fig. 13.1 Typical TEM image of carbon colloids obtained

20 nm

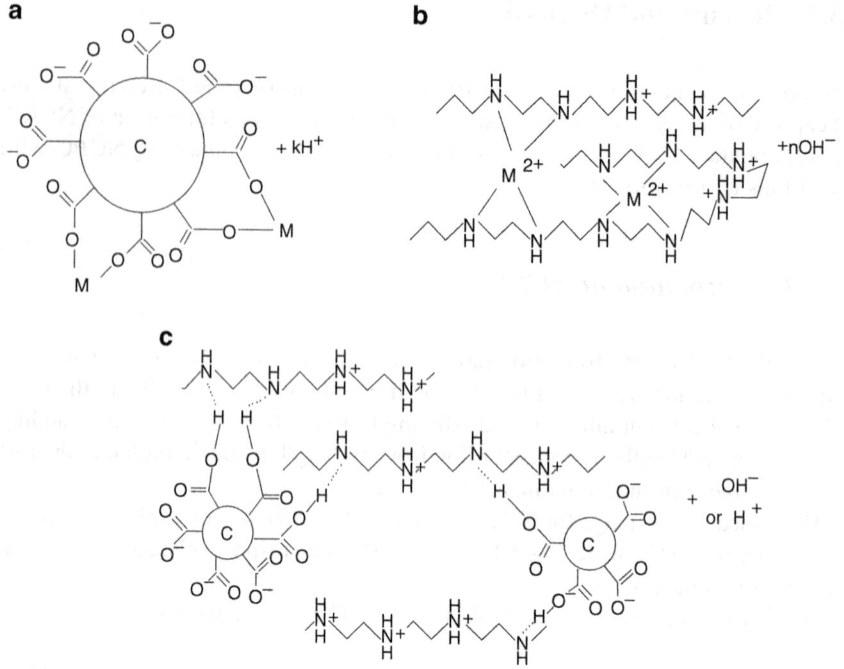

a

b

c

Fig. 13.2 The structures of (**a**) Carbon nanoparticle with an attached metal ion, (**b**) polymeric complex PEI-metal ions and (**c**) the formation of NCPC

NCC with carboxyl groups on its surface interacts with divalent metal ions. Figure 13.2a demonstrates schematically a carbon nanoparticle with attached metal ion:

$$\text{Carb} - 2\text{COO}^- - \text{M}^{2+} \tag{13.5}$$

In aqueous solution PEI combines with a proton according to the following equilibrium reaction:

$$\text{PEI} + n\,\text{H}_2\text{O} \leftrightarrow \text{PEI H}_n^{\,n+} + n\,\text{OH}^- \tag{13.6}$$

Where $0 \leq n \leq n_m$ (n_m – the number of monomers -CH$_2$-CH$_2$-NH- contained in a single polymeric chain) and depends on pH of the solution.

The mechanism of PEI interaction with metal ions M^{2+} can be described by the following equilibrium reaction:

$$\text{PEI} + a\,\text{M}^{2+} \leftrightarrow \text{PEI M}_a^{\,2a+} \tag{13.7}$$

Where $0 \leq a \leq a_m$, $a_m = n_m/4$. The idealized structure of polymeric complex PEI-metal ions is given in Fig. 13.2b.

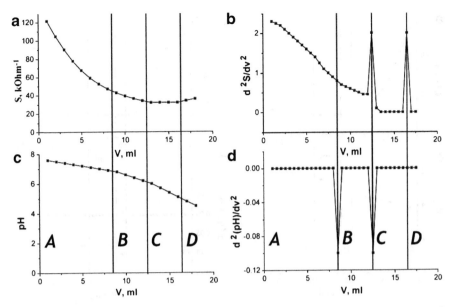

Fig. 13.3 Dependences of conductivity S and pH of mixed solution with initial concentration of PEI of 100 mg/l against volume V of NCC solution with concentration of 58 mg/l which is added to PEI solution (part **a** and **c**) and second-order derivatives of these curves (part **b** and **d**)

The reaction of NCC with PEI in water solution leads to formation of NCPC (Fig. 13.2c). The conductivity and pH values of mixed solutions were measured to study the process of NCPC formation. Figure 13.3a, c demonstrate the dependence of conductivity S and pH of mixed solution with initial concentration of PEI of 100 mg/l against the volume V of NCC solution with concentration of 58 mg/l which is added to PEI solution. Dependences of conductivity S and pH of mixed solution with initial concentration of NCC of 80 mg/l against the volume V of adding PEI solution with concentration of 4 mg/l are given in Fig. 13.4a, c. Second-order derivatives of these curves were calculated and are presented in Figs. 13.3b, d and 13.4b, d. Temperature of solutions in these and described below experiments was 25°C.

When NCC is added to PEI solution, the conductivity S and pH of the solution decrease (cf. Fig. 13.3, interval AB). In the interval BC not all of the carboxyl groups can react with PEI because of the influence of charges of neighbour chains of PEI, therefore the excess of the number of carboxyl groups is required. Coagulation process begins with a progressive acceleration from the point C and reaches the maximal speed in the point D, whilst the conductivity S does not change within the interval CD.

Figure 13.4b demonstrates the process of coagulation. After point D when all NCPC precipitated, conductivity S of the solution increases due to increasing the concentration of H$^+$ - ions in the solution.

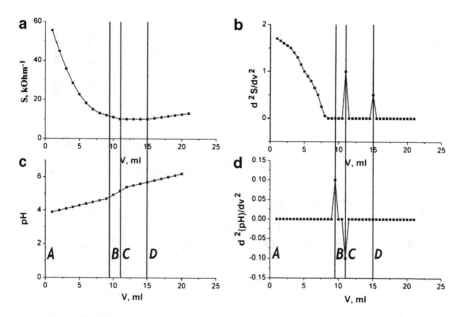

Fig. 13.4 Dependences of conductivity S and pH of mixed solution with initial concentration of NCC of 80 mg/l against volume V of adding PEI solution with concentration of 4 mg/l (part **a** and **c**) and second-order derivatives of these curves (part **b** and **d**)

When PEI solution is added to NCC, conductivity S of solution decreases and pH increases (Fig. 13.4, interval AB). The interval BC corresponds to the event when the excess of the number of carboxyl groups are required to react with PEI. Coagulation process begins with progressive acceleration from the point C and reaches the maximal speed in the point D; the conductivity S does not change in the interval CD. After point D when all NCPC precipitated, the conductivity S of the solution increases due to increasing the concentration of OH^- - ions in the solution.

Experiments have shown that at the point D of Figs. 13.3 and 13.4 when NCPC precipitates, the ratio of concentrations of PEI C_{pei} and NCC C_{ncc} does not depend on the concentration of PEI in the interval from $2 \cdot 10^{-4}$ to 0.1%. But the specific surface of NCC increases with decreasing their diameter and the ratio $(C_{pei}/C_{ncc})_D$ depends on the mean size of NCC (Fig. 13.5). These results show that the use of NCC with small sizes is more effective. For example, for NCC with size of 20 nm the ratio $(C_{pei}/C_{ncc})_D = 0.42 \pm 0.07$ and $(C_{pei}/C_{ncc})_D = 6.7 \pm 1.1$ for NCC with the mean size of 5 nm.

The speed of NCPC coagulation process versus the ratio of concentrations of PEI and NCC C_{pei}/C_{ncc} was investigated. Obtained results for NCC with the mean size of 20 nm are given in Fig. 13.6 and show that the coagulation time within the interval CD is very short and dramatically increases with decreasing of NCC concentration. This feature can be effectively used in methods of removing metal ions from water and remediation of contaminated soil.

Fig. 13.5 Dependence of the ratio $(C_{pei}/C_{ncc})_D$ from the mean size of NCC

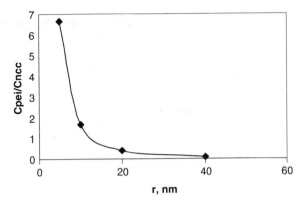

Fig. 13.6 The speed of NCPC coagulation process versus the ratio of concentrations of PEI and NCC C_{pei}/C_{ncc}

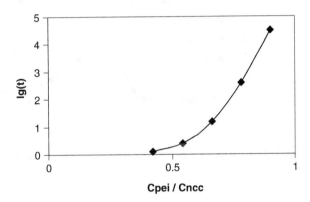

13.3.2 *Removing Metal Ions by NCPC from Water*

As it was mentioned above the process of metal ions removal from water solutions involves an interaction of metal ions with PEI and NCC. Simultaneously the positively charged complex M - PEI combines with carboxyl groups of NCC with the NCPC formation as follows: (i) by hydrogen bond between carboxyl group of NCC and amine group of PEI which have not formed coordination bonds with M^{2+}, (ii) by electrostatic interaction between negatively charged carboxyl ion of NCC and positively charged M-PEI complex. After the interval of time depending on the ratio C_{pei}/C_{ncc} the NCPC coagulates and can be easily removed by filtration or centrifugation.

The ability of the method to remove metal ions Zn (II), Cd (II), Cu (II), Hg (II), Ni (II), and Cr (VI) were investigated. In these tests the mean size of NCC was 20 nm, the ratio C_{pei}/C_{ncc} was 0.75, PEI and metal ions concentrations were 150 mg/l and 1 mg/l, respectively. Figures 13.7 and 13.8 represents the dependence of the removal ratio P against pH of solutions containing metal ions with concentration of

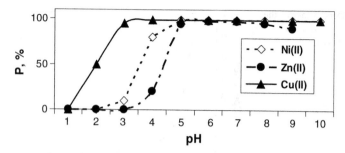

Fig. 13.7 The dependence of the removal ratio P for Cu (II), Zn (II) and Ni (II) against pH of solutions

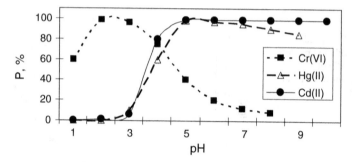

Fig. 13.8 The dependence of the removal ratio P for Cd (II), Hg (II) and Cr (VI) against pH of solutions

Table 13.2 Bonding capacity Q and distribution coefficients K_d

Parameters	Cr(VI)	Ni(II)	Cu(II)	Zn(II)	Cd(II)	Hg(II)
Q, mmol/g	2.0 (pH=6) 4.0 (pH=2)	5.5	5.7	5.2	5.0	4.4
K_d	13 (pH=6) 1,600 (pH=2)	1,100	1,200	990	760	35

10 mg/l. Distribution coefficients K_d at pH=6 are given in Table 13.2. The bonding capacities Q were obtained from adsorption isotherms of metal ions at pH=6, V=50 ml, W=7.5 mg, temperature t=25°C and at the contact time of 5 min are given in Table 13.2.

Time courses of metal ions adsorption were conducted using solutions of 10 mg/l concentration (Figs. 13.9 and 13.10). The amount of metals adsorption increased rapidly during beginning of 30–40 s (about 90% removal). Subsequently, the adsorption rate rises gradually and reaches equilibrium (about 100% removal) after 50–70 s. The short time required to reach equilibrium implies that the NCPC have very high adsorption efficiency and have a great potential in divalent metals adsorbent application.

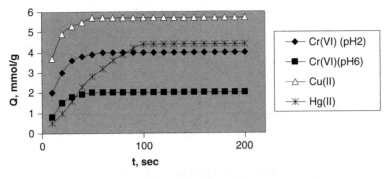

Fig. 13.9   Adsorption of Cr (VI), Cu (II) and Hg (II) metal ions vs process time

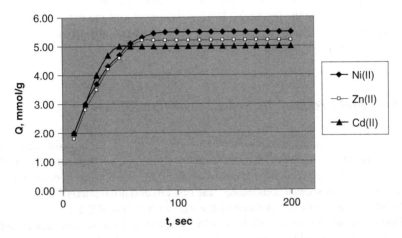

Fig. 13.10   Adsorption of Ni (II), Zn (II) and Cd (II) metal ions vs process time

13.3.3   Potentiality of NCPC in Water Treatment

Laboratory tests were conducted to reveal the efficiency of NCPC for purification of contaminated water. For this purpose 200 l. enamel container with a drainage pipe and a discharge valve at the bottom was connected through a water pump with a sand filter. The enamel container was filled with water containing metal ions Cd^{2+}, Cu^{2+}, Zn^{2+} with concentrations of 3, 10 and 5 mg/l, respectively; temperature of the water was 24°C, pH =6.8. 4 l of 0.5% NCPC solution with the ratio $C_{pol}/C_{ncc} = 0.78 \pm 0.15$ was prepared and added in the container with contaminated water. The water in the container was mixed thoroughly during 2 min and in 6 min after that the solution was discharged through the valve, water pump and sand filter. The flow rate was about 1,000 l/h and it was adjusted by the valve. Three samples of the water flowing out the sand filter were taken every 4 min after beginning the water

Table 13.3 Test results of proposed water purification method

Metal ions	Initial concentration, mg/l	Concentration in water flowing out the sand filter, µg/l		
		After 4 min	After 8 min	After 12 min
Cu	10	5.1 ± 0.3	5.0 ± 0.3	4.9 ± 0.3
Zn	5	10.2 ± 0.5	9.8 ± 0.5	10.3 ± 0.5
Cd	3	5.3 ± 0.3	5.1 ± 0.3	5.2 ± 0.3
PEI	0.044	< 0.5	< 0.5	< 0.5

discharge process to determine concentrations of Cd, Cu and Zn. Test results given in Table 13.3 confirm that: (i) the life time of NCPC with the ratio $C_{pei}/C_{ncc} = 0.78 \pm 0.15$ is not greater than 8 min, (ii) the sediment after NCPC coagulation is easily removed even by a sand filtration, (iii) the filtered water does not contain NCPC and PEI (at least at lower than the detectable levels); (iv) metal ions are not washed off the sediment during the filtration process. Thus the obtained results demonstrate a high efficiency of the proposed water purification method.

13.3.4   Field Tests of the Soil Remediation Method

Field tests were conducted to reveal the efficiency of NCPC for remediation of soil and groundwater at the place of former tannery where soil was contaminated by Cr (VI). The ground surface with the area of about 80 m^2 was divided in equal parts to compare characteristics of soils treated (the first lot) and not treated (the second lot) by NCPC. Each of lots was isolated by the wall with height of 0.2 m.

First of all, an identity of soils in both lots was tested. Ground core samples with diameter of 20 mm and length of 70 mm were taken from 5 different places of each lot from the depth of 0–0.07, 0.2–0.27, 0.4–0.47, 0.6–0.67 and 0.75–0.82 m. Small stones were removed from the samples, and then each sample was mixed thoroughly, weighed, dried at 110°C during 15 min and weighed again to determine the moisture of samples. 100 mg of the each sample was taken to determine Cr concentration C_{Cr} by NAA. Mean value of humidity and C_{Cr} of soil of each depth value and lot were calculated. Then each sample was put into a retort and was filled with 200 ml of water and kept for 48 h. After that the water was filtered through a paper filter and the concentration of extracted Cr was determined in water. The results of these tests are given in Table 13.4.

As one can see from Table 13.4, C_{Cr} in the contaminated soil is about 60 mg/kg in both lots and the contamination extends for a depth over 0.8 m. Moreover Cr concentration in extracted water is high, about 200 µg/l. The soil moisture is about 3% in both of the lots. These results show that the lots are identical and one of them can be used as the experimental lot and another as the reference one.

In order to evaluate the efficiency of the proposed method of soil remediation, we have treated the first lot by NCPC. 80 l of 0.5% NCPC solution with the ratio

Table 13.4 Comparison of soil characteristics of 2 lots

Parameter	Depth, m				
	0–0.07	0.2–0.27	0.4–0.47	0.6–0.67	0.75–0.82
	The first lot				
C_{Cr} in soil, mg/kg	68±20	54±16	61±18	50±15	46±15
Extracted C_{Cr} in water, µg/l	230±50	200±40	210±40	190±40	165±35
Soil moisture, %	4±1	3±1	3±1	3±1	3±1
	The second lot				
C_{Cr} in soil, mg/kg	64±20	50±16	69±18	60±15	54±15
Extracted C_{Cr} in water, µg/l	200±50	230±40	230±40	210±40	185±35
Soil moisture, %	4±1	3±1	3±1	3±1	3±1

Table 13.5 Comparison of soil characteristics of lots after treatment by NCPC

Parameter	Depth, m				
	0–0.07	0.2–0.27	0.4–0.47	0.6–0.67	0.75–0.82
	The first lot				
C_{Cr} in soil, mg/kg	62±20	59±16	58±18	60±15	50±15
Extracted C_{Cr} in water, µg/l	1.3±0.2	1.5±0.2	1.4±0.2	190±40	170±35
Soil moisture, %	11±3	13±3	11±3	10±3	3±1
	The second lot				
C_{Cr} in soil, mg/kg	60±20	63±16	63±18	57±15	54±15
Extracted C_{Cr} in water, µg/l	230±50	220±40	200±40	180±40	190±35
Soil moisture, %	12±3	11±3	12±3	11±3	3±1

C_{pei}/C_{ncc} =0.85 was prepared. This concentrated solution was used to prepare the working solution with concentration of NCPC of 0.01%. The first lot was treated during 8 h by 4 m³ of working solution. The second lot was just watered without NCPC. Ground core samples from these two lots were taken from 5 different places and depths, treated and tested as described above. Test results are given in Table 13.5.

As shown in Table 13.5, the value of C_{Cr} in contaminated soil was not changed after watering. Over t=8 h the water penetrated into the soil up to depth of L_w=0.7 m. The soil moisture deeper than 0.7 m was not changed after watering. The Cr concentration in extracting waters of ground core samples taken from depths up to L_{NCPC}=0.5 m is very low, about 1.4 µg/l, but it rises sharply from the depth of 0.5 m. It means that the life time t_{NCPC} of NCPC before coagulation

$$t_{NCPC} = L_{NCPC} / L_W \cdot t \qquad (13.8)$$

Is about 5.5–6 h and NCPC penetrated up to depth of 0.5 m. This value of NCPC life time is in a good accordance with the data given in Fig. 13.6.

The obtained results demonstrate a high efficiency of the proposed soil remediation method. The depth or area of remediation depends on NCPC life, i.e., on the value of ratio C_{pei}/C_{ncc} and the penetration rate of NCPC solution. Accordingly, the depth of penetration of NCPC in soil or depth of remediation of soil can change from 1 to 100 cm, and distance of NCPC moving with groundwater or remediation zone of the ground can change from 1 to 100 m.

13.4 Conclusion

The method of obtaining nanocarbon - polymer nanocomposites on the base of nano-carbon colloids and polyethylenimine has been studied. Nanocarbon- polyethylenimine nanocomposites are formed due to carboxyl groups on the surface of carbon nanoparticles.

Life time of NCPC in the interval between points B and C is very long, not less than 1 year. Life time of NCPC decreases after point C. Its value depends on the ratio of concentrations of polymeric molecules and carbon nanoparticles and lies in interval 1 s to 1,000 min.

The position of point D when NCPC precipitates does not depend on the concentration of PEI in the interval from $2 \cdot 10^{-4}$ to 0.1% but depends on ratio $(C_{pei}/C_{ncc})_D$. In-turn, the ratio $(C_{pei}/C_{ncc})_D$ depends on the mean size of NCC because the specific surface of NCC depends on their diameter. For example, for NCC with size of 20 nm the ratio $(C_{pei}/C_{ncc})_D = 0.42 \pm 0.07$ and $(C_{pei}/C_{ncc})_D = 6.7 \pm 1.1$ for NCC with the mean size of 5 nm.

The metal ions interact with NCPC via ion exchange and complexation mechanism. Studies of NCPC sorption properties show that the composition has high bonding capacity of 4.0–5.7 mmol/g at pH = 6 for most of divalent metal ions such as Zn (II), Cd (II), Cu (II), Hg (II), Ni (II), Cr (VI), distribution coefficients of ions are 10^1–10^3. The obtained results show that NCPC have very high adsorption efficiency and have a great potential in divalent metals adsorbent application.

The process of metal ions removal from water solutions involves the following stages: (i) sorption of metal ions by PEI and NCC with simultaneous formation of NCPC, (ii) coagulation of NCPC containing metal ions and (iii) removing the coagulated NCPC by filtration or centrifugation in order to recover the metals. The contaminated soil can be remediated due to the coagulation of NCPC forming insoluble compounds with metal ions.

Laboratory tests have shown that when water is contaminated by metal ions at the level of 1–10 mg/l, the described method allows 100–1,000 times decreasing the concentration of metal ions in water; the sediment after NCPC coagulation is easily removed by filtration; the filtered water does not content NCPC and PEI (at least lower than the detectable level). The obtained results demonstrate a high efficiency of the proposed water purification method.

The depth of penetration of NCPC in a soil or the depth of remediation of soil can vary from 1 to 100 cm, and the distance of NCPC moving with groundwater or the remediation zone of ground can vary from 1 to 100 m. Field tests have shown that when soil is contaminated by Cr (VI) at the level of 50–60 mg/kg, the described method allows 10^2 times decrease of concentration of metal ions in water passing through the soil.

References

1. Abderrahim O, Amine Didi M, Moreau B, Villemin D (2006) A new sorbent for selective separation of metal: polyethylenimine methylenephosphonic acid. Solvent Extr Ion Exc 24(6):943–955
2. Chen C, Wang X (2006) Adsorption of Ni (II) from aqueous solution using oxidized multiwall carbon nanotubes. Ind Eng Chem Res 45:9144–9149
3. Hsu WK, Terrones M, Hare JP, Terrones H, Kroto HW, Walton D (1996) Electrolytic formation of carbon nanostructures. Chem Phys Lett 262:161–166
4. Hudson MJ, Hunter-Fujita FR, Pecketta JW, Smithb PM (1997) Electrochemically prepared colloidal, oxidised graphite. J Mater Chem 7(2):301–305
5. Juang RS, Chiou CH (2000) Ultra filtration rejection of dissolved ions using various weakly basic water-soluble polymers. J Membr Sci 177:207–214
6. Juang RS, Chiou CH (2001) Feasibility of the use of polymerassisted membrane filtration for brackish water softening. J Membr Sci 187:119–127
7. Juang RS, Shiau RC (2000) Metal removal from aqueous solutions using chitosan-enhanced membrane filtration. J Membr Sci 165:159–167
8. Khaydarov RR, Khaydarov RA, Gapurova O (2009) Remediation of contaminated groundwater using nano-carbon colloids, nanomaterials: Risk and benefits, NATO science for peace and security series C: environmental security. Springer, Dordrecht, pp 219–224
9. Khaydarov RA, Khaydarov RR, Gapurova O (2010) Water purification from metal ions using carbon nanoparticle-conjugated polymer nanocomposites. Water Res 44:1927–1933
10. Kim D, Hwang Y, Cheong SI, Lee JK, Hong D, Moon S, Lee JE, Kim SH Production and characterization of carbon nano colloid via one-step electrochemical method. J Nanopart Res. doi:10.1007/s11051-008-9359-2
11. Li YH, Di Z, Ding J, Wu D, Luan Z, Zhu Y (2005) Adsorption thermodynamic, kinetic and desorption studies of Pb2+ on carbon nanotubes. Water Res 39:605–609
12. Li YH, Ding J, Luan Z, Di Z, Zhu Y, Xu C, Wu D, Wei B (2003) Competitive adsorption of Pb2+, Cu2+ and Cd2+ ions from aqueous solutions by multi-walled carbon nanotubes. Carbon 41:2787–2792
13. Li YH, Wang S, Luan Z, Ding J, Xu C, Wu D (2003) Adsorption of cadmium (II) from aqueous solution by surface oxidized carbon nanotubes. Carbon 41:1057–1062
14. Li YH, Wang S, Wei J, Zhang X, Xu C, Luan Z, Wu D, Wei B (2002) Lead adsorption on carbon nanotubes. Chem Phys Lett 357:263–266
15. Li YH, Zhu Y, Zhao Y, Wu D, Luan Z (2006) Different morphologies of carbon nanotubes effect on the lead removal from aqueous solution. Diamond Relat Mater 15:90–94
16. Liang P, Liu Y, Guo L, Zeng J (2004) Multi-walled carbon nanotubes as solid-phase extraction adsorbent for the pre-concentration of trace metal ions and their determination by inductively coupled plasma atomic emission spectrometry. J Anal At Spectrum 19:1489–1492
17. Lu C, Chiu H (2006) Adsorption of zinc (II) from water with purified carbon nanotubes. Chem Eng Sci 61:1138–1145
18. Lu C, Chiu H, Bai H (2007) Comparisons of adsorbent cost for the removal zinc (II) from aqueous solution by carbon nanotubes and activated carbon. J Nanosci Nanotechnol 7:1647–1652
19. Lu C, Chiu H, Liu C (2006) Removal of zinc (II) from aqueous solution by purified carbon nanotubes: kinetics and equilibrium studies. Ind Eng Chem Res 45:2850–2855
20. Lu C, Liu C (2006) Removal of nickel (II) from aqueous solution by carbon nanotubes. J Chem Technol Biotechnol 81:1932–1940
21. Molinari R, Gallo S, Pietro Argurio P (2004) Metal ions removal from wastewater or washing water from contaminated soil by ultra filtration–complexation. Water Res 38:593–600
22. Peckett JW, Trens P, Gougeon RD, Poppl A, Harris RK, Hudson MJ (2000) Electrochemically oxidised graphite .Characterization and some ion exchange properties. Carbon 38:345–353
23. Rao GP, Lu C, Su F (2007) Sorption of divalent metal ions from aqueous solution by carbon nanotubes. A rev Sep Purif Technol 58:224–231

24. Rumeau M, Persin F, Sciers V, Persin M, Sarrazin J (1992) Separation by coupling ultrafiltration and complexation of metallic species with industrial water soluble polymers. Application for removal or concentration of metallic cations. J Membr Sci 73:313–322
25. Steenkamp GC, Keizer K, Neomagus H, Krieg H (2002) Copper (II) removal from polluted water with alumina/chitosan composite membranes. J Membr Sci 197:147–156
26. Vieira M, Tavares CR, Bergamasco R, Petrus JCC (2001) Application of ultra filtration-complexation process for metal removal from pulp and paper industry wastewater. J Membr Sci 194:273–276

Chapter 14
Aspects for Execution and Finalisation of Groundwater Remediation Measures

Joerg Frauenstein, Jochen Grossmann, and Joerg Drangmeister

Abstract According to the German Federal Soil Protection Act, the soil, contaminated sites, and any water pollution caused by harmful soil changes shall be remediated in such a manner that no hazards, considerable disadvantages or considerable nuisances for individuals or the general public occur long term.

Prior to implementation of any measure, a remediation investigation regulated by Annex 3 to the Federal Soil Protection and Contaminated Sites Ordinance (BBodSchV) is prescribed in the form of a comparative review of suitable measures (e.g. remediation methods and strategies). The stipulated measure and its consequences for the polluter must be in reasonable proportion to the hazard which has to be prevented. This means that preference must be given to that measure/combination of measures which, while being equally effective, represents the "milder means" (i.e. is necessary) and which exhibits an adequate cost-benefit ratio.

Due to the complex circumstances involved in each individual case of contamination (such as geological and hydrogeological site characteristics, specific nature of the impact, and relevance of the protected assets affected by specific uses), no thresholds have been legally prescribed under German law for determining the need for remediation, nor have remediation target values been defined. Instead, the competent authorities were accorded a considerable degree of discretion, which has proved its worth in enforcement.

J. Frauenstein (✉) • J. Grossmann
Section: Soil Protection Measures, Federal Environment Agency, Woerlitzer
Platz 1, Dessau-Rosslau 06844, Germany
e-mail: joerg.frauenstein@uba.de

J. Drangmeister
GICON – Großmann Ingenieur Consult GmbH, Tiergartenstraße 48, 01219
Dresden, Germany

F.F. Quercia and D. Vidojevic (eds.), *Clean Soil and Safe Water*, NATO Science for Peace and Security Series C: Environmental Security, DOI 10.1007/978-94-007-2240-8_14, © Springer Science+Business Media B.V. 2012

14.1   Introduction

Groundwater and drinking water standards are very much aligned in Germany due to significant usage in this regard. Drinking water quality must meet high standards. The Drinking Water Ordinance [5], which transposed the 1998 EC Drinking Water Directive into national law, prescribes these standards. Some of the basic requirements stipulate that drinking water must not only be free of pathogens and substances in concentrations that may be harmful to health, it must also be "wholesome and clean". Therefore a good chemical as well as a good ecological status is required [2].

However, more than 80% of contaminated sites in Germany have a significant impact on groundwater. Therefore, groundwater remediation is a serious task within contaminated land management and in fact plays a leading role among remediation measures. Furthermore, groundwater remediation is in practice the main cost driver. In order to contribute to economically and ecologically acceptable solutions to the contaminated sites problem, further development of criteria and strategies for remediating groundwater degradation and innovative remediation strategies (with a special focus on megasites) are required to ensure the usage of groundwater for drinking water purposes.

Consequently, the need to analyze and evaluate any groundwater remediation measure in line with the temporal and quantitative development of the contamination source and plume, the achievement of remediation targets within aquifers, and the use of state-of-the-art groundwater remediation (in terms of technology and installations) is clear. Therefore, the complex system of soils and groundwater has to be taken into account. In practice, we are faced with several sources of errors and mistakes in the decision processes. In addition, we have to cope with a lack of available and reliable data and information. Nevertheless, we dispose about multiple corrective actions within the system as we keep the discrepancies between theory and practice in mind. Such regulating screws exist especially during planning and preparation of a measure, with a proper site specific investigation, with single case related assessment and prognosis, with an appropriate process understanding, with analogous observations among similar cases, and last but not least with a state-of-the-art implementation of science and technology regarding degradation and/or remediation.

Further corrective actions before and during groundwater remediation measures exist in the form of the configuration and ideal position of the clean-up installation (in particular for the remediation wells) and the consideration of secondary processes caused by the remediation itself. In addition, the optimization of the operational mode of the clean-up installation is a requisite to ensure effectiveness of the treatment. If all means fail, only an adjustment of remediation targets helps. In practice, the responsible authorities must often decide on termination criteria for the remediation measures.

In the case that remediation targets will surely be reached, this is a routine decision. However, more realistic is a resulting delay in combination with disproportionate effort to achieve the last few percentages of the remediation target.

14.2   Retrospective Analysis of Finalized Groundwater Remediation

As part of the Environmental Research Program of the Federal Environment Ministry, a project was set up with GICON GmbH, in Dresden [1]. The objective was to develop an approach to evaluate the effectiveness of groundwater remediation and a method to assess results which could be legally used by the environmental authorities. Therefore, the project was conducted in cooperation with 12 German Federal states.

The objective of the project was to develop a technical and legal basis for the development of a concept for dealing with existing groundwater contamination.

Therefore, in particular:

- the temporal and quantitative development of pollutant spread,
- the availability of remediation targets in the aquifer, and
- the state of technology (process or system) based on active and passive remediation and groundwater protection measures were researched and evaluated.

Furthermore, the effects of a series of impacting factors on the course, the results, the costs, and ongoing restructuring measures were carried out and assessed for the transport of contaminants in groundwater.

This should provide guidance for planning, implementation, and completion of groundwater remediation and to increase the operational and economic efficiency of groundwater remediation in practice.

The result of the project should provide the first systematic approach for proper handling of groundwater contamination cases in practice, and should also allow prognoses on the progress of groundwater remediation.

The foundation of the research was a case book for groundwater impacts and their rehabilitation, which was provided by the participating States.

The project was divided into the following phases:

- **Phase 1**: Creation of a questionnaire and "GWKON" database programming – Based on this questionnaire for the assessment of groundwater remediation cases, a comprehensive database system with user-friendly interfaces and analysis routines was developed. Figures 14.1 and 14.2 shows the development from initial concept to extensive database structure to construct all relevant interdependencies.
- **Phase 2**: Data collection from 96 (finalized) groundwater remediation measures – After the programming of "GWKON" and the presentation of the functions and requirements, the selection of appropriate remediation cases, the data preparation and input into the database system was compiled by participating States.
- **Phase 3**: Appraisal of case studies and creation of a guidance document – The results of data analysis were aggregated by the program into graphs and tables. These provided an overview of stored data as a basis for further evaluation.

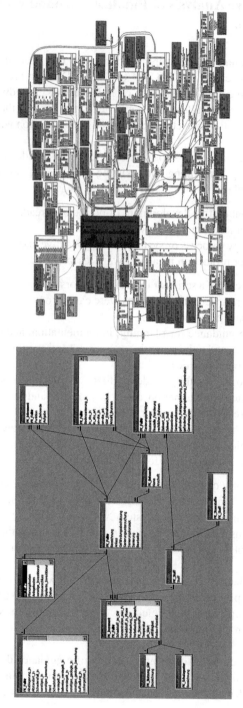

Fig. 14.1 Database structure and interdependencies from initial concept to final implementation

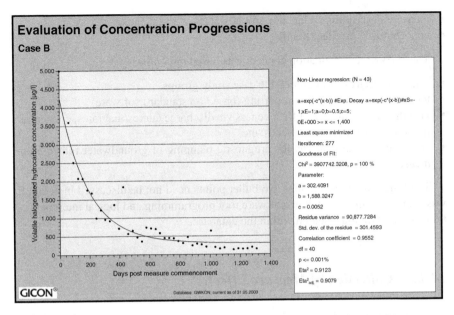

Fig. 14.2 Evaluation of concentration discharge data

In an additional step, the results were evaluated in order to ensure implementation of the conclusions and a guideline document for legally-compliant use of groundwater contamination cases.

14.3 Evaluation of the Case Collection

14.3.1 General Implementation

To learn more about the prevailing interrelation among the case studies, a complete data set is preferable. Therefore, the data base is structured in a way to collect as much information as possible regarding the following significant topics:

- general information
- site-specific data
- geological and hydrogeological information
- general groundwater information
- impact symptoms
- status of protected objects
- measure descriptions related to soil

- measure descriptions related to groundwater
- concentration profiles (trend)

The "GWKON" evaluation concept was based on 4 pillars:

- statistical analysis of comparable case study groups
- tool box for searching and reporting (based on standard software)
- visualization of measure effectiveness via discharge curves/diagrams, concentration output, time and cost trends, etc.
- "potential" considerations for prognostic planning of groundwater remediation measures

The functionality of the first two bullet points need not be discussed in detail. It was primarily implemented by software tool programming and logical interconnections for an individual search within the database.

14.3.2   Collection of Measure Effects

For well-documented cases of groundwater impact, curves of remediation discharge (with conversion into concentration output) were determined and statistically evaluated.

The concentration changes in the aquifer (in-situ concentrations) were similar to the evaluated and presented discharge curves. The ratios of discharge were then evaluated for groundwater concentration based on the correlation curves. This allows a scatter-independent delineation. The discharge concentration at the beginning of a remediation action is usually not dense enough. Therefore, a compensation of errors due to substantial changes in concentration is possible (Figs. 14.2 and 14.3). A stable level of discharge concentration among 80% of in-situ measured cases after 3 years of remediation could be displayed.

Without existence of on-site measurements, this represents a very rough estimation. However, it offers the possibility of comparison between existing and potential discharge and also among cases.

For more detailed investigations of any potential hydrogeological differentiable impacted unit, isoclines maps, data on soil mechanics and soil chemical characteristics are required. At that point, potential shifts due to hydraulic effects can be more precisely compiled. This must also consider the thickness distribution. On this basis, a measured evaluation and optimization of individual cases is also possible.

14.3.3   Observations from the Case Study Evaluation

In most case studies, a complete data set did not exist, although we only used finalised remediation data. Thus, some obligatory data are indicated, but without a case study

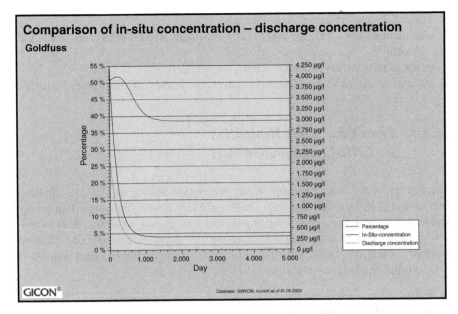

Fig. 14.3 Comparison between in situ-concentration and concentration discharge

it cannot be used for further aggregation. In fact, the "GWKON" database includes 89 groundwater remediation measures. After checking completeness and plausibility, only 67 cases could be utilized for further investigation. Seventy-five percent of the cases involved chlorinated hydrocarbons (CHC) as the dominant contaminant, in 16 cases PAH / BETX was identified as the main pollutant and more than 80 out of 89 cases utilized pump and treat measures. From the present case collection (subject to consideration of correlations) statistically reliable statements for CHC cases only in combination with pump and treat can be derived. However, for 15 cases, a markedly unique characteristic dominates and does not permit any cross-case analysis.

An intention of the project was the search for a new approach to achieve comparability among cases of the same pollutant group and to find a simplified tool for the derivation of characteristic values for the effectiveness of remedial measures. Ultimately, the question had to be answered whether the chance to transfer single case-based results into an overarching prognosis approach existed.

The basis for such a determination includes reliable details on the pre-remediation groundwater damage, the actual discharge of contaminants during remediation, and the development of pollutant concentrations caused by the measure itself. These values are time-dependent and can be visualized by discharge and concentration curves. This case collection does not necessarily give a representative view of groundwater remediation practice in Germany. However, it provides indications as to deficiencies in practical implementation, such as:

- incomplete data sets for any further aggregation
- knowledge on the extent of the damage was often uncertain
- an existing dominance of the individual case

- uncertainties in the remediation forecast
- uncertainties about existing legal issues and likely needed efforts and costs for remediation
- a lack of information exchange among stakeholders
- incomplete and inadequate documentation of case studies and about crucial criteria

14.3.4   Cross Case Study Evaluation
for CHC-Contaminated Sites

The next objective was to determine any correlation among similar case studies. Due to available data, this was only feasible for CHC-contamination within pump and treat measures. For overarching criteria and a prognoses, a balance sheet model was created. This tool attempted to forecast the effect of theoretically-relevant boundary conditions on the remediation progress within the contaminant source.

The following boundary conditions were implemented:

- surface impacts (immission)
- thickness of unsaturated zone
- thickness of saturated zone
- CHC contamination in saturated zone groundwater
- permeability of the aquifer (k_f)
- pore volume of the aquifer
- groundwater gradient
- groundwater recharge

Regarding the contamination situation:

- contamination of groundwater flow (upstream)
- spread of contamination into a plume
- soil contamination in an unsaturated zone
- soil contamination in a saturated zone

From these figures, simplified water balances, emission potentials and associated rates are calculated. Thus, we attempt to predict the interaction between the hydraulic measure and the potential trend. The potential trend can be reconverted into concentration figures.

With these results, it is possible to predict concentration profiles within the source zone (in-situ concentration), the pollutant concentration already removed or still to be removed from groundwater, and the remaining potential in affected zones of soil and groundwater.

The major determinant of removable pollution potential (taken out by the hydraulic measure) is the concentration balance between groundwater in retention areas and in usable pore spaces. The desorption of particle-bound pollutants in soil particle potential in the retention area and the usable pore space is also a determining factor for the remaining potential.

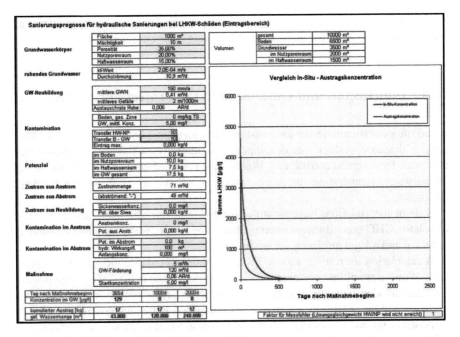

Fig. 14.4 Calculation spreadsheet (balance model)

Fundamentally known effect relationships to describe the solution equilibrium (e.g. the function of the concentration gradient), were already implemented in the approach. However, we still require so-called "transfer factors" as a control variable to simulate the impact of these potential transitions in remediation measures.

This theory can be tested with real cases and operational scenarios from the case studies as a type of general plausibility check for the model. Moreover, the transfer factors for the transition from potential pollution coming from the pore space or from the soil particles into groundwater can be determined recursively in terms of a simplified adaptation.

Nevertheless, it must be noted that this forecast delivers idealized, homogeneous, "average" conditions over the entire area of impacted groundwater. This only allows for the evaluation of generalized, parameterized statistical data (Fig. 14.4).

The result is a "typical" or idealized curve for certain boundary conditions. For this purpose, it is absolutely necessary to consider complete information and data concerning the case. This is the only way to review the all-determining transfer factors and curves, and to adjust in line with actual measured data.

Detailed forecasts are only possible with appropriate flow and contaminant transport modelling, which requires a very high level of detail. The proposed tool cannot be a replacement for modelling, but it is sufficient enough to discuss a remediation measure, and to provide evidence for the optimization of hydraulic measures against CHC contamination.

Generally, it makes complex geo-hydraulic and geochemical interactions visible.

Various constellations and, in particular, the effect of the sensitivity of different parameters for remediation are comparable among different scenarios.

14.4 Conclusions and Outlook from the Project Perspective

Based on a questionnaire for executed groundwater remediation cases, a comprehensive database system was developed. The majority of the case studies were processed by assigned consultants. Eighty-nine recorded cases of mostly finalised groundwater remediation measures were stored in a database for further evaluation.

Statistical correlations were identified for hydraulic remediation (pump and treat) of volatile CHC groundwater contamination. The results were implemented into a technical and a legal guidance document.

A calculation instrument was developed based on a balance model. This instrument allows a simplified presentation of effects during the course of a groundwater remediation. Consequently, different cases are comparable with regard to the constellation and the sensitivity of parameters and effects within a scenario.

Based on the balance model, a first assumption as to the efficiency and the achievable success of the CHC remediation in groundwater is possible. The emission curves and the concentration trend of hydraulically-remediated CHC contamination showed that the interactions of soil-mechanical properties defining the hydraulic dynamics in the groundwater, and the regime of the groundwater withdrawal determine the efficiency, the course and finally the expense of a measure.

Further, it was demonstrated that residual contamination in soil, being in the saturated or in the unsaturated zone, hinder the achievement of remediation targets within a timeframe of up to 15 years, even in case of lower contamination potentials.

These interrelations derived from the study require an approach deviating from present practice.

These apply to:

- the steps of investigating a contamination
- the risk assessment
- the identification of remediation goals and consequent derivation of remediation target values
- the planning of measures and their implementation
- the decision on termination of remedial measures

Still in focus are the following questions which derive from the study outcome:

- Which criteria determine the progress and the success of remediation measures?
- Can scenarios predict the course of remediation?

14.5   Legal Requirements Regarding Soil and Groundwater Protection

Soils and contaminated sites, and any water pollution caused by harmful soil changes or contaminated sites, shall be remediated in such a manner that no hazards, considerable disadvantages or considerable nuisances for individuals or the general public occur long term.

The remediation investigation regulated by Annex 3 to BBodSchV is a comparative review of suitable measures (e.g. remediation methods and strategies). In this review, the principle of proportionality must be observed: the measures stipulated by the competent authority, and their consequences for the party obligated to carry them out, must be in reasonable proportion to the hazard to be prevented. This means that preference must be given to that measure / combination of measures which, while being equally effective, represents the "milder means" (i.e., is necessary) and which exhibits an adequate cost-benefit ratio.

Due to the complexity of the circumstances of each individual case (such as geological and hydrogeological site characteristics, specific nature of the impact, relevance of the protected assets affected by specific uses), no legal thresholds have been prescribed for determining the need for remediation, nor have remediation target values been defined. Instead, the competent authorities were accorded a considerable degree of discretion, which has proved its worth in enforcement. However, in practise we have identified the following legal and technical deficits:

- reliable legal interpretation (groundwater impacts, remediation targets)
- court-resistant criteria for proportionality analysis
- a better understanding of the soil-groundwater system
- tried and tested alternatives (innovative) remediation measures with sufficient practical experience
- experienced and well-trained personnel from all parties involved
- suitable instruments to predict the course of a remediation measure

From the authorities involved in the execution, the following was in particular required:

- criteria for or against a remediation decision
- recommendations for trustworthy investigations and ability tests
- additional assessment criteria for the stipulation of remediation targets
- assistance during design, implementation and monitoring of groundwater remediation measures
- nature and extent of stakeholder participation, legal assessment of the conformity of their recommendations
- setting up criteria for the termination of (active) groundwater remediation

14.6   Ongoing Activities to Modify Existing Regulations in Germany

Due to advances in science and technology and the experiences gained in enforcement, the secondary legislation needs to be updated. The German Federal Environment Ministry detailed necessary adjustments and updates in a draft proposal for amendment to the BBodSchV:

- fundamental revision of Annex 1 to BBodSchV (requirements concerning sampling, methods of analysis and quality assurance during the investigations): [4]

  - update all references to standards and establish investigational methods for new priority pollutants
  - update the methods for estimating substance inputs to groundwater from suspect sites and sites suspected of being contaminated
  - determination of the equivalence of analytical methods in laboratory practice and dealing with measurement uncertainty in enforcement

- updating and supplementing the trigger and action values listed in Annex 2 to BBodSchV
- harmonising the trigger values for evaluation of the soil-groundwater pathway with the LAWA marginal thresholds, and formulation of implementing rules (In 2004, the Joint Water Commission of the States (Länderarbeitsgemeinschaft Wasser, LAWA) derived so-called marginal thresholds (Geringfügigkeitsschwellenwerte, GFS) for 71 individual substances and summative parameters. The marginal threshold defines the boundary between an insignificant change in chemical groundwater quality and harmful contamination.)
- taking natural attenuation into consideration when deciding on remedial, protective and restrictive measures

Future fields of action are:

- summarize findings for further contaminants for the optimization of groundwater remediation
- develop innovative remediation strategies (with a special focus on megasites)
- further develop criteria and strategies for remediating groundwater degradation due to contaminated sites
- gain experience with in-situ technologies and other promising measures
- implementation of monitored natural attenuation concepts via a guideline for authorities "Consideration of natural attenuation in remediating contaminated sites" [5]

References

1. Drangmeister J, Großmann J, Willand A (2007) Cross-national criteria for the management of groundwater contaminations, UBA Nr. 20/2007, download: http://www.umweltdaten.de/publikationen/fpdf-l/3202.pdf (German, with an English abstract)
2. Federal Ministry for the Environment, Nature conservation and nuclear safety (BMU) (2010) Water framework directive – the way towards healthy waters. Download: http://www.bmu.de/files/pdfs/allgemein/application/pdf/broschuere_wasserrahmenrichtlinie_en_bf.pdf
3. Frauenstein J (2010) Current state and future prospects of remedial soil protection, download: http://www.umweltdaten.de/publikationen/fpdf-l/4041.pdf
4. Kabardin B, Frauenstein J (eds) (2011) Consideration of natural attenuation in remediating contaminated sites, download: http://www.umweltbundesamt.de
5. TrinkwV (2001) http://www.gesetze-im-internet.de/bundesrecht/trinkwv_2001/gesamt.pdf

References

Chapter 15
Advances in Groundwater Remediation: Achieving Effective *In Situ* Delivery of Chemical Oxidants and Amendments

Robert L. Siegrist, Michelle Crimi, Mette M. Broholm, John E. McCray, Tissa H. Illangasekare, and Poul L. Bjerg

Abstract Contamination of soil and groundwater by organic chemicals represents a major environmental problem in urban areas throughout the United States and other industrialized nations. *In situ* chemical oxidation (ISCO) has emerged as one of several viable methods for remediation of organically contaminated sites. Many of the most prevalent organic contaminants of concern at sites in urban areas (e.g., chlorinated solvents, motor and heating fuels) can be destroyed using catalyzed hydrogen peroxide (H_2O_2), potassium permanganate ($KMnO_4$), sodium persulfate ($Na_2S_2O_8$), or ozone (O_3) delivered into the subsurface using injection wells, probes, or other techniques. A continuing challenge for ISCO, as well as other *in situ* remediation technologies, is how to achieve *in situ* delivery and obtain simultaneous contact between treatment fluids, such as oxidants and amendments, and the target contaminants. During the past few years, advances have been made in several key areas including knowledge and know-how associated with: (1) use of amendments for enhanced delivery and distribution of treatment fluids in heterogeneous settings with zones of low permeability media, (2) use of direct push technology for targeted high resolution delivery of treatment fluids, and (3) use of monitoring and sensing methods for direct feedback for delivery control and evaluation of remediation effectiveness. This paper provides a summary of ISCO and highlights ongoing

R.L. Siegrist (✉) • J.E. McCray • T.H. Illangasekare
Environmental Science and Engineering, Colorado School of Mines, 112 Coolbaugh Hall, Golden, CO 80401, USA
e-mail: siegrist@mines.edu

M. Crimi
Institute for a Sustainable Environment, Clarkson University, 8 Clarkson Avenue, Potsdam, NY 13676-1402, USA

M.M. Broholm • P.L. Bjerg
Department of Environmental Engineering, Technical University of Denmark, Anker Engelunds Vej 1, 2800 Kgs, Lyngby, Denmark

F.F. Quercia and D. Vidojevic (eds.), *Clean Soil and Safe Water*, NATO Science for Peace and Security Series C: Environmental Security, DOI 10.1007/978-94-007-2240-8_15, © Springer Science+Business Media B.V. 2012

efforts to advance the effective *in situ* delivery of treatment fluids, with an emphasis on chemical oxidants and amendments, which can help achieve cleanup goals and protect groundwater and associated drinking water resources.

15.1 Contaminated Sites and Remediation Methods

Contamination of soil and groundwater is present at many sites within or near urban areas throughout the United States and other countries. These sites include gasoline stations, drycleaners, manufacturing facilities, power plants, mining operations, waste disposal facilities, and military installations. At these sites, organic chemicals are typically the primary contaminants of concern (COCs) in soil and groundwater. At gasoline stations and sites with underground storage tanks for fuels, the primary COCs are volatile organic compounds (VOCs) including benzene, toluene, ethyl-benzene, and xylenes (BTEX). At sites with industrial activities, chlorinated solvents are often present including trichloroethene (TCE), perchloroethene (PCE), chloroform, and carbon tetrachloride. Chlorinated solvents can be present as dense nonaqeuous phase liquids (DNAPLs) and present a long-term problem due to their chemical properties and stability in the environment, which can contribute to high contaminant concentrations in groundwater over extensive time frames. Similarly, light nonaqueous phase liquids (LNAPLs; e.g., gasoline and fuel products) in the subsurface can also be sources of long-term groundwater contamination.

Contamination of soil and groundwater by organic chemicals represents a major environmental problem in urban areas throughout the United States and other industrialized nations [33, 38]. Over many decades, a wide variety of toxic organic chemicals have intentionally or accidentally been released into the subsurface resulting in serious risks to human health and environmental quality (e.g., increased cancer risk through ingestion of contaminated drinking water or inhalation of vapours within buildings).

Early attempts to clean up contaminated sites commonly involved using pumping wells to extract contaminated groundwater so it could be treated aboveground and discharged to the surface (i.e., pump-and-treat). Over time, it became apparent that this approach had severe performance limitations and high costs and that cleanup of contaminated groundwater using pump-and-treat alone was virtually impossible [16]. As a result, research and development efforts were focused on finding cost-effective *in situ* technologies based on both engineered and natural attenuation processes (Fig. 15.1). In addition, remediation of difficult sites was increasingly viewed as best accomplished by combining remedies simultaneously or sequentially for different zones of contamination. For example, treatment of a DNAPL source zone with an aggressive technology like thermally enhanced extraction or *in situ* chemical oxidation (ISCO) might enable remediation of the associated groundwater plume using engineered bioremediation or permeable reactive barriers.

ISCO is one of several technologies that have emerged with the potential for cost effective remediation of contaminated soil and groundwater. ISCO involves

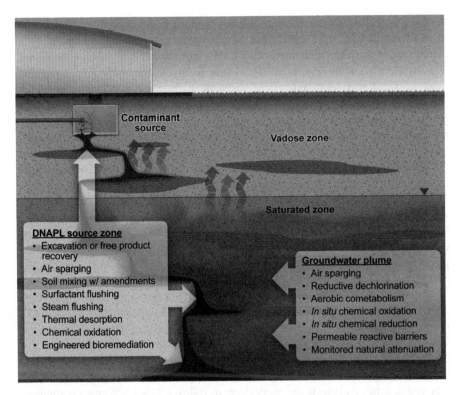

Fig. 15.1 Illustration of *in situ* technologies and approaches developed for remediation of sites contaminated by organic chemicals from Siegrist et al. [33]

the subsurface delivery of a chemical oxidant to destroy organic COCs and thereby reduce potential risks to public health and environmental quality.

The current chemical oxidants in relatively widespread use for ISCO are hydrogen peroxide (H_2O_2), potassium and sodium permanganate ($KMnO_4$, $NaMnO_4$), sodium persulfate ($Na_2S_2O_8$), and ozone (O_3) [17, 32, 33].

15.2 *In Situ* Chemical Oxidation for Remediation of Contaminated Sites

The principles and practices of applying ISCO for remediation of soil and groundwater contamination are thoroughly described in several recent review papers and reference books (e.g., [13, 14, 17, 33, 40]). Thus only a brief overview is provided here.

Many organic chemicals can be destroyed through chemical reactions with oxidants, such as catalyzed hydrogen peroxide (also known as CHP or modified

Fenton's reagent), potassium and sodium permanganate, sodium persulfate, or ozone [31, 33]. Under the right conditions, oxidants generate reactive species that can transform and often mineralize many COCs including chlorinated hydrocarbons (e.g., TCE, PCE), fuels (e.g., benzene, toluene, methyl tertiary butyl ether), phenols (e.g., pentachlorophenol), polycyclic aromatic hydrocarbons (e.g., naphthalene, phenanthrene), polychlorinated biphenyls, explosives (e.g., trinitrotoluene), and pesticides (e.g., lindane). Degradation reactions tend to involve electron transfer or free radical processes with simple to complex pathways with various intermediates, and typically follows second-order kinetics. The need for activation to generate free radicals and the sensitivity to matrix conditions, such as temperature, pH, and salinity, vary with the different oxidants and specific contaminants.

An oxidant and amendments (if needed) can be delivered into a target treatment zone (TTZ) within the subsurface at varied concentrations and mass loading rates in liquid, gas, or solid phases. Delivery has most commonly been accomplished through permeation by vertical direct-push injection probes or flushing by vertical groundwater wells. Other delivery approaches have employed horizontal wells, infiltration galleries, soil mixing, and hydraulic or pneumatic fracturing.

Krembs et al. [17] developed a database of ISCO project applications and completed a critical review of field application practices and experiences. Of 242 ISCO projects examined, PCE or TCE were the targeted COCs at 70% of the sites, the subsurface conditions were characterized as permeable at 75% of the sites, and oxidants were delivered using permanent or temporary injection wells at 70% of the sites. For 99 full-scale ISCO projects that attempted to meet a specific goal and reported that they had met it, the results were as follows: 21% of 28 projects attempting to achieve drinking water maximum contaminant limits (MCLs) met this goal; 44% of 25 projects attempting to achieve alternative concentration limits (ACLs) met this goal; 33% of 6 projects attempting to reduce the COC mass by a certain percentage met this goal; 82% of 34 projects attempting to reduce the COC mass and/or time to cleanup met this goal; and 100% of 6 projects attempting to evaluate effectiveness and optimize future injections met this goal.

Consistent with the observations Krembs et al. [17], the effectiveness of ISCO depends on COC properties, site conditions and remediation objectives. At some sites, ISCO has been applied and the destruction of the target COCs has occurred such that cleanup goals have been met in a cost-effective and timely manner. At other sites, ISCO applications have had uncertain or poor in situ treatment performance. Poor performance has often been attributed to poor uniformity of oxidant delivery caused by low permeability zones and formation heterogeneity, excessive oxidant consumption by natural subsurface materials, or presence of large masses of DNAPLs.

ISCO system selection, design, and implementation practices should rely on a clear understanding of ISCO and its applicability to a given set of contaminant and site conditions to achieve site-specific remediation objectives. A number of key issues may be relevant and need to be addressed regardless of the oxidant and delivery system being employed, including: (1) amenability of the target COCs to oxidative degradation; (2) effectiveness of the oxidant for NAPL destruction; (3) optimal oxidant loading (dose concentration and delivery) for a given target treatment zone in a given subsurface setting; (4) non-productive oxidant consumption due to the natural oxidant demand (NOD) exerted by natural organic matter, reduced inorganic species, and some mineral phases in the subsurface; (5) non-productive oxidant consumption due to autodecomposition reactions and free radical scavenging reactions; (6) potential adverse effects (e.g., mobilizing metals such as chromium, forming toxic by-products, reducing formation permeability, generating off-gases and heat); and (7) potential to combine ISCO with other remediation technologies and approaches [33]. Thus, the site-specific application of ISCO requires consideration of a variety of questions and careful attention to key points to enable success and avoid problems (Table 15.1).

Table 15.1 Key points to consider, which can enable success with ISCO and help avoid problems (Adapted from Siegrist et al. [33])

Key points to keep in mind for *In Situ* Chemical Oxidation

1. *ISCO has great potential for successful use at some, but not all sites.* ISCO is very applicable to sites with relatively small TTZs where there are permeable subsurface conditions and low to moderate mass levels of COCs. ISCO can successfully achieve common treatment goals at these types of sites.

2. *ISCO can successfully achieve treatment goals.* Treatment goals need to be realistically set and based upon site-specific conditions and challenging circumstances.

3. *Effective in situ delivery is essential. In situ* delivery methods must be matched to the oxidant and site conditions to achieve ISCO treatment goals.

4. *ISCO will normally require two or more active delivery events.* To achieve effective distribution of oxidant throughout a TTZ, and the essential contact of oxidant and target COCs, normally two or more delivery events will be required in all of, or just portions of, the TTZ.

5. *One or more oxidants can destroy most, if not all, of the common organic COCs.* Reaction rates and the extent of destruction are not limiting, if a compatible oxidant is used for the target COCs and if they are brought into contact for enough time to achieve treatment.

6. *ISCO can be, and often must be, synergized with other remedies.* ISCO can be used in combined remedies for synergistic outcomes. A prime example of a combined remedy involves ISCO followed by enhanced reductive dechlorination or monitored natural attenuation.

7. *So-called "rebound" will often occur, more so at some sites than others.* Rebound is defined as an increase in the concentrations of COCs in groundwater after active ISCO operations have ended. Rebound can be useful to help refine the conceptual site model and to optimize the design of follow-up treatments.

8. *ISCO can temporarily perturb subsurface conditions.* ISCO can change conditions within a TTZ (e.g., depress pH, mobilize certain redox-sensitive metals, decrease the activity of certain microbes), but effects are normally not long lasting or of consequence in most settings.

9. *Oxidant transport by diffusion is often negligible.* Oxidants with slower reaction rates (e.g., permanganate or persulfate) have an ability to penetrate lower permeability media (LPM) and help mitigate back diffusion from LPM. However, the rate of oxidant diffusion into these LPM zones is typically extremely slow compared to the rate of reaction.

10. *The cost of ISCO varies widely.* The cost of an ISCO depends on various factors. For example, sites with fuel hydrocarbons and permeable subsurface conditions typically cost less while those with DNAPLs or complex subsurface conditions typically cost more.

15.3 Continuing Challenges for ISCO and Other *In Situ* Technologies

A continuing challenge for ISCO, as well as other *in situ* remediation technologies, is how to achieve *in situ* delivery and obtain simultaneous contact between oxidants and amendments and the target COCs. During the past few years, advances have been made in several key areas including knowledge and know-how associated with: (1) use of amendments for enhanced delivery and distribution of treatment fluids in heterogeneous settings with zones of low permeability media, (2) use of direct push technology for targeted high resolution delivery of treatment fluids, and (3) use of monitoring and sensing methods for direct feedback for delivery control and evaluation of remediation effectiveness. These advances have the potential to improve groundwater remediation based on more effective *in situ* delivery of treatment fluids – including oxidants and amendments – to remediate a TTZ and achieve cleanup goals that help protect groundwater and associated drinking water resources. Highlights of research and development work recently completed and ongoing at the Colorado School of Mines, Clarkson University, and The Technical University of Denmark are provided below.

15.3.1 Amendments for Enhanced Delivery

Advances are occurring in the use of amendments for enhanced delivery and distribution of treatment agents in heterogeneous settings with low permeability zones (e.g., [10, 35, 36]).

Poor treatment performance for remediation technologies involving fluid delivery into the subsurface is often attributed to inadequate uniformity of *in situ* distribution (also known as poor sweep efficiency) caused by zones of low permeability media (LPM). When injected under an applied pressure gradient the resulting subsurface distribution of treatment fluids is impacted greatly by the architecture of the subsurface permeability field because oxidants and other treatment amendments seek preferential flow paths through more permeable media, resulting in a less efficient sweep of the target treatment zone by the injected fluid. This leads to the injected fluid bypassing LPM, which can result in rebounding of contaminant concentrations within a groundwater zone following cessation of active delivery. The extent to which this occurs in a given heterogeneous system largely depends on the physicochemical properties of the injected fluid, the mode of introduction (e.g., injection rates, orientation and placement of well screens), the permeability distribution, the location of the target treatment zone (e.g., in high-permeability zones, within clay zones, etc.), and the interaction of the fluid with the solid media at the pore-scale. Therefore, understanding the interplay between the site-specific heterogeneity of the subsurface and the injected remediation fluids is crucial to optimizing the distribution of applied amendments in the subsurface, thereby enhancing the contact between the amendment and the target contaminant.

Mobility control methods, including a class of strategies involving the modification of *in situ* fluid viscosities, exist that have the potential to mitigate the effects of permeability heterogeneity. A fundamental laboratory and mathematical modelling study of the applicability of heterogeneity control for groundwater remediation in shallow hydrogeologic systems was recently completed [23]. This study focused on a viscosity-modification strategy developed by the petroleum industry for enhanced oil recovery to overcome preferential flow and other bypassing effects produced by large-scale geological heterogeneities. Traditional mobility control techniques in petroleum reservoir engineering have involved the use of polymers that increase the viscosity of the injected solutions. The increased viscosity of the injected fluid minimizes the effects of the aquifer heterogeneities by promoting strong transverse fluid movement, or cross-flow, across heterogeneous reservoir units [18, 37], providing an enhanced sweep efficiency. The occurrence and benefits of cross-flow during polymer flooding for oil recovery is well documented (see [30, 37] and references therein) and a summary of recent applications in environmental restoration may be found in Jackson et al. [15]. However, the applicability of this technology for site-scale heterogeneities that are important in groundwater remediation, particularly during co-application with biological and chemical remediation amendments has not been previously investigated.

Specific to ISCO, xanthan gum was identified as a highly promising polymer for use with permanganate oxidant solutions because it maintains a stable and predictable viscosity within the oxidant solution and exhibits a low oxidant demand for permanganate [35]. Xanthan also was shown to be non-toxic to two different strains of PCE degrading bacteria, as well as a general microbial consortium used in beer-brewing; and was deemed as a possible electron donor for PCE degrading bacteria [23, 36]. Thus it may also provide a useful technique to facilitate biodegradation.

When the xanthan gum/permanganate solution contacts PCE (aqueous or non-aqueous phase liquid), solution viscosity rapidly decreases. This coupled with the low oxidant demand for the polymer suggests that the oxidation of PCE initiates partial oxidation of the xanthan molecule at specific locations along the polymer chain, impacting the viscosity. However, as a part of design, during subsurface injection a continuous and stable bank of xanthan/permanganate solution will exist behind the PCE contact zone that will continue to impart heterogeneity control within the aquifer. An example of polymer-improved sweep efficiency in a 2D experimental tank is presented in Fig. 15.2 [23, 34]. For these 2D experiments, sands with different grain sizes were carefully packed in layers to yield physical heterogeneity structures with permeability's between layers varying by 100× or more. As illustrated in Fig. 15.2, the addition of 800 ppm of xanthan gum can provide enhanced sweep efficiency (i.e., swept volume of the tank divided by the total tank volume). Sweep efficiencies can be enhanced by a factor of two after only one pore volume (PV) of throughput of an aqueous tracer solution containing xanthan gum biopolymer compared to the tracer solution alone. Furthermore, for the lowest permeability layers, the sweep efficiency can be increased by a factor of five or more when xanthan gum is used. Numerical simulations demonstrated that the permeability contrast as well as the order of layering was important in governing sweep efficiency [23, 34].

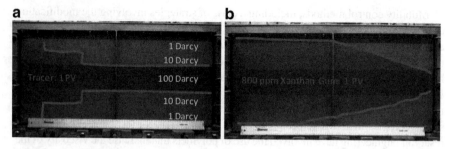

Fig. 15.2 2D tank experiment with a fivelayer sand heterogeneity structure illustrating delivery of water containing a conservative solute tracer alone (**a**) and with addition of xanthan gum biopolymer (**b**) (from McCray et al. [23] and Silva [34]). Note: the lines within the tanks reveal the boundaries of the water delivered along the left side of each 2D tank after 1 PV of throughput

It is useful to note that even modest permeability contrasts (e.g., factor of 2–5) can cause significant flow bypassing in contaminated aquifers. Thus, these experimental results suggest that the addition of polymers to remediation amendments can greatly improve amendment delivery efficiency into relatively lower permeability strata and reduce the volume of amendment required to achieve such delivery.

In addition to difficulties due to naturally existing site heterogeneities, MnO_2 particles, a product of permanganate ISCO, can create secondary site heterogeneities that may provide an added hindrance to the ISCO technology's effectiveness. MnO_2 particles have the potential to deposit in the well and subsurface and impact the flow regime in and around the permanganate injection system, including the well screen, filter pack, and the surrounding subsurface formation. Permeability changes resulting from MnO_2 particle deposition have been observed under some conditions during laboratory and field evaluations (e.g., [11, 19, 20, 22, 28, 42, 43]).

Crimi and Ko [8] determined that MnO_2 particles in groundwater can be controlled using amendments to specifically allow for their facilitated transport through porous media. The ideal amendment interacts minimally with porous media, reacts minimally with the oxidant permanganate, interacts minimally with other groundwater components, is acceptable to the regulatory community, and is cost-effective. Four potentially viable amendments were evaluated in their research under a range of geochemical conditions (pH, ionic content and concentration, particle concentration, and oxidation-reduction potential). Particle behaviour was evaluated using spectrophotometric methods and particle size was measured using filtration and optical measurements. Sodium hexametaphosphate (SHMP) was found to best meet the qualifications of the ideal amendment. When SHMP is included in a permanganate oxidation solution, the resulting MnO_2 particles were sub-micron in size and remained suspended in solution (i.e., mobile) under all of the conditions evaluated during batch tests. Additional experiments to evaluate the transport of MnO_2 both with and without SHMP in 1-D transport systems of varied porous media content (i.e., organic matter, clay, mineralogy) revealed SHMP enhanced MnO_2 stability during transport through porous media [9]. It is noted that SHMP may not be

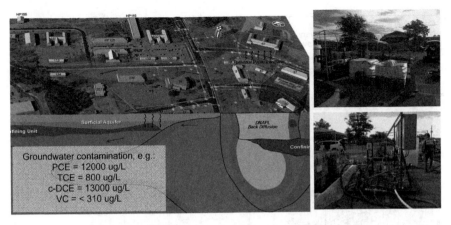

Fig. 15.3 Field demonstration of ISCO for remediation of chlorinated solvent contaminated groundwater using co-injection of oxidant and amendments to enhance delivery (from Crimi et al. [10]). The ISCO system includes co-injection of $KMnO_4$ (5 g/L), SHMP (5 g/L), and Xanthan gum (0.5 g/L) during a single well injection event to treat a 10m diameter zone of sand and silty sand in the depth interval of 11 to 16m below ground surface

applicable to porous media with high contents of expandable clays due to potential for clay dispersion and permeability loss.

Currently, Crimi et al. [10] are conducting a field demonstration to evaluate the combined use of xanthan and SHMP to overcome the effects of site heterogeneity through improved uniformity and extent of oxidant delivery. The demonstration is ongoing at Marine Corps Base Camp Lejeune near Jacksonville, North Carolina, USA (Fig. 15.3). The base covers approximately 236 square miles and is a training base for the United States Marine Corps. The test area is located within the site of the former base dry cleaning facility, which operated from the 1940s until 2004 when the building was demolished. The subsurface in the test area is comprised of sands of varying permeability with some interbedded silt layers. PCE contamination has been detected at high concentrations in soil and groundwater samples (Fig. 15.3).

15.3.2   Delivery Enhancement Technologies

Advances continue in the use of methods for targeted high-resolution delivery of oxidants and amendments (e.g., [4–7, 12, 29]).

Three methods for enhanced delivery of amendments for *in situ* remediation of low-permeability deposits were tested at a field site in Denmark by Christiansen et al. [5, 7]: (1) pneumatic fracturing, (2) direct-push delivery, and (3) hydraulic fracturing. The three methods are illustrated in Fig. 15.4. The field site was a former distribution center for organic chemicals, where the handling and storage of chemi-

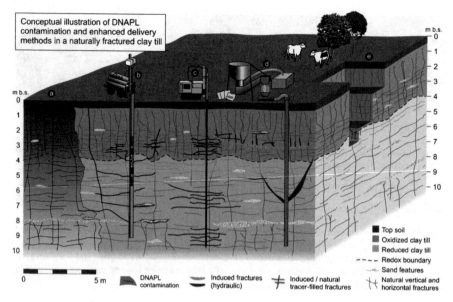

Fig. 15.4 Illustration of results of a field demonstration involving side-by-side tests of direct-push technology, pneumatic fracturing, and hydraulic fracturing in a clay till in Denmark. A multi-component tracer mixture was composed of brilliant blue, fluorescein, and Rhodamine WT tracers and the pathway spacing, thickness and radius of influence were assessed through direct documentation via coring and excavation with sample collection and tracer analyses ([6])

cals had led to severe contamination of the subsurface, primarily with chlorinated solvents. The tests were conducted at an uncontaminated part of a site in a basal clay till, which based on thorough geological characterization was deemed a geologically representative basal clay till site, at testing depths of 2.5–9.5 m below ground surface (bgs). For all three of the delivery methods tested, the tracer brilliant blue and the fluorescent tracer's fluorescein and Rhodamine WT were delivered to allow for visual detection under normal and UV light. The pathway spacing, thickness and radius of influence of the delivered solution were assessed through direct documentation via coring and excavation with sample collection and tracer analyses. An overview of the distribution of tracers from each injection method is illustrated in Fig. 15.4.

Pneumatic fracturing with nitrogen gas applied with delivery intervals of 1 m (see in Fig. 15.4b) resulted in tracer spreading in a dense networks of natural fractures with a radius of influence of less than 2 m above the redox boundary (0–3 m bgs) and in sub-horizontal discrete and widely spaced, induced tracer-filled fractures at depths >3 m bgs. Denser spacing may be obtainable if shorter delivery intervals are applied. Venting of the gas and tracer to the surface (or unsaturated zone) occurred when the induced fractures short-circuited in natural sub-vertical fractures or boreholes.

Direct-push delivery with a Geoprobe® (see in Fig. 15.4c) was conducted for every 10–25 cm vertical spaced depth in three depth-intervals in a single point and in a cluster of three points. This resulted in distributed tracer primarily in natural fractures above the redox boundary and in discrete, closely spaced (but not merging) induced fractures below the redox boundary with a radius of influence of about 1 m. The direct-push delivery method was robust and efficient for enhanced delivery at the clay till site.

Hydraulic fracturing with a sand-guar mixture at 3 m bgs (see in Fig. 15.4d) produced an elliptical, asymmetrical, bowl-shaped fracture with a radius of approximately 3.5 m (green shape at 3 m depth bgs in of Fig. 15.4d). The geometry hydraulic fractures induced at 6.5 m bgs (red shape at 6.5 m depth in of Fig. 15.4d) and 9.5 m bgs is uncertain, but clearly not horizontal. This method appears more influenced by *in situ* stress conditions than the other methods. Therefore, geotechnical tests are recommended.

Overall, the direct-push delivery method was most successful as a close vertical spacing of delivery points was obtained without significant merging. The technology is flexible and relatively low cost. Based on the field demonstrations completed, the direct push method is recommended for enhanced amendment delivery at contaminated basal clay till sites.

15.3.3 Monitoring and Sensing Plume Response

Monitoring the remediation of a TTZ often involves collection of samples of soil and aquifer solids for analysis of target contaminants. The time and cost required for such sampling and analysis can be high and errors and uncertainties in the data can be quite large (e.g., [16, 25]). An alternative approach is to evaluate the response of a groundwater plume during and following remediation of a TTZ (e.g., [25]). This is normally done using samples collected at a set of monitoring wells which are analyzed for target contaminants and by-products associated with a specific remediation approach. The samples often have to be sent to specialized testing laboratories for analysis. As a result, in cases where groundwater plume response has to be evaluated over long time periods, this approach for monitoring can often be cost prohibitive.

An alternative to this conventional approach to monitoring is based on sensors and new advances in wireless data transmission. This alternative provides an "intelligent" way of collecting and processing large amounts of data and has the potential to be used for low-cost monitoring during remediation of groundwater over long periods of time. As an example, combined remediation schemes are evolving where aggressive mass removal is followed by a passive approach such as monitored natural attenuation. In these situations, the monitoring during passive remediation can be done effectively using automated sensor based technologies where data could be collected and transmitted for long-term monitoring.

A set of sensors with the capability to measure chemical concentrations can be installed at sampling points and connected to miniature computing devices (known as motes) to form a wireless sensor network (WSN). A mote consists of a microprocessor, memory, sensors, analog-to-digital converters, a data transceiver, and energy source (usually battery and/or solar) and controllers integrating these components. With these capabilities, the motes have the capability to collect data from the

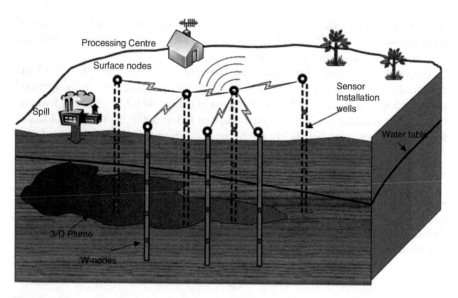

Fig. 15.5 Illustration of the use of an integrated, subsurface chemical plume monitoring system, capable of capturing transient plumes in real-time to assess the source and treatment effectiveness

sensors and deliver that data to a central site (computer) using custom network protocols. The motes can conduct the necessary computations to process data locally and form ad hoc networks of sensor nodes that get re-configured as the plume configuration changes due to spreading and/or remediation. Figure 15.5 shows schematically how a WSN is deployed in the field.

WSN have been used to monitor active volcanoes [41]. An early application of sensor networks for subsurface monitoring was reported by Ramanathan et al. [27] to detect arsenic in rice fields in Bangladesh. Musaloiu et al. [24] reports on a study where temperature, moisture, light, and CO_2 levels were monitored to study a soil ecosystem. Akyildiz and Stuntebeck [1], after evaluating the viability of using the WSN technology in the subsurface, concluded that many challenges have to be overcome because the radio signals are impacted by soil. Trubilowicz et al. [39] reported on their experience in deploying a WSN in a watershed and suggested that WSN will be a very useful tool for field studies but a number of advances have to be made on both hardware and software.

As a first step in developing the WSN technology for subsurface plume monitoring applications, a proof-of-concept study was conducted by Porta et al. [26] in a two-dimensional experimental aquifer. A two-dimensional intermediate scale tank was packed with sands to represent a heterogeneous unconfined aquifer. A set of ten electrical conductivity sensors interfaced with motes formed the network. Communication software that was developed for data acquisition was tested. Transient plumes were created using NaBr tracer. The lessons learned from the proof-of-concept were used to upscale the system to a three-dimensional experimental aquifer (Fig. 15.6). This 3D aquifer has dimensions of 4.9 m long × 2.4 m

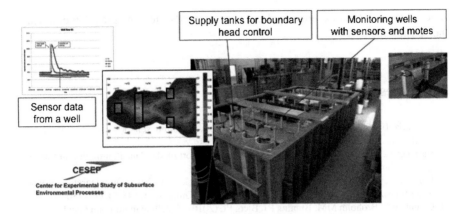

Fig. 15.6 Illustration of a 3D test bed for experimental studies of integrated monitoring with wireless sensor networks

high × 0.6 m wide (16 ft × 8 ft × 2 ft). Multi-compartment end reservoirs are used to create different groundwater flow configurations. Tracers are introduced into the end reservoirs to create solute plumes. No-flow blocks were placed in the tank to divide the plume to create conditions so that a WSN could capture the dynamics of bifurcating plumes representing the plume responding to remediation. The WSN consists of 36 sensors interfaced with motes. The data generated was used to test fault detection algorithms [3, 21] and inversion methods [2].

The proof-of-concept studies and the subsequent findings suggest that for WSN technology to be adopted for ground water plume monitoring, a number of issues have to be resolved. These include signal transmission issues in subsurface environments, energy efficiency of the system, sensor types and data accuracy and failure of sensors. Additional research and development are needed to address these to make WSN a viable technology.

15.4 Summary

In situ remediation has advanced during the past decade and there is a growing recognition of the potential benefits of combining multiple remediation technologies and approaches to improve delivery and contaminant destruction or removal and using integrated approaches for monitoring remediation progress and assessing cleanup effectiveness. *In situ* chemical oxidation is one of a few robust and reliable methods for groundwater remediation. While the principles and practices are well understood and established, advancements are still needed to improve effectiveness at sites with contamination in heterogeneous subsurface conditions including those with low permeability zones. Advances are occurring, including those highlighted in this paper: co-injection of stabilization aids and polymers, direct-push and fracture-enhanced delivery methods, and integrated monitoring and sensing methods.

References

1. Akyildiz IF, Stuntebeck EP (2006) Wireless underground sensor networks: Research challenges. Ad Hoc Netw 4:669–686
2. Barnhart K, Illangasekare TH (2011) Automatic transport model data assimilation in Laplace space. Submitted to Water Resour Res
3. Barnhart K, Urteaga I, Han Q, Jayasumana A, Illangasekare T (2010) On integrating groundwater transport models with wireless sensor networks. Groundwater 48(5):771–780
4. Chambon C, Broholm MM, Binning PJ, Bjerg PL (2010) Modelling multi-component transport and enhanced anaerobic dechlorination processes in a single fracture – clay matrix system. J Contam Hydrol 112:77–90
5. Christiansen CM, Riis C, Christensen SB, Broholm MM, Christensen AG, Klint KES, Wood JSA, Bauer-Gottwein P, Bjerg PL (2008) Characterization and quantification of pneumatic fracturing effects at clay till site. Environ Sci Technol 42:570–576
6. Christiansen CM (2010) Methods for enhanced delivery of *in situ* remediation amendments in contaminated clay till. PhD thesis. DTU Environment, Technical University of Denmark, Kgs Lyngby, Denmark, ISBN 978-87-91855-87-0
7. Christiansen CM, Damgaard I, Broholm MM, Kessler T, Klint KE, Nilsson B, Bjerg PL (2010) Comparison of delivery methods for enhanced *in situ* remediation in clay till. Ground Water Monit R 30(4):107–122
8. Crimi M, Ko S (2009) Control of manganese dioxide particles resulting from *in situ* chemical oxidation using permanganate. Chemosphere 74(6):847–853
9. Crimi M, Quickel M, Ko S (2009) Enhanced permanganate *in situ* chemical oxidation through MnO_2 particle stabilization: Evaluation in 1D transport systems. J Contam Hydrol 105(1–2):69–79
10. Crimi M, Silva JAK, Palaia T (2011) Cooperative technology demonstration: polymer-enhanced subsurface delivery and distribution of permanganate. ESTCP Project no. ER-0912 http://www.serdp.org/
11. Heiderscheidt J, Siegrist RL, Illangasekare TH (2008) Intermediate-scale 2D experimental investigation of *in situ* chemical oxidation using potassium permanganate for remediation of complex DNAPL source zones. J Contam Hydrol 102(1–2):3–16
12. Hønning J, Broholm M, Bjerg P (2007) Role of diffusion in chemical oxidation of PCE in a dual permeability system. Environ Sci Technol 41:8426–8432
13. Huling SG, Pivetz BE (2006) Engineering Issue: In-Situ Chemical Oxidation. U.S. Environmental Protection Agency, 600-R-06-702
14. ITRC (2005) Technical and regulatory guidance for *in situ* chemical oxidation of contaminated soil and groundwater, 2nd Edn. (ISCO-2). The Interstate Technology & Regulatory Council *In Situ* Chemical Oxidation Team. www.itrcweb.org/gd_ISCO.asp
15. Jackson RE, Dwarakanath V, Meinardus HW, Young CM (2003) Mobility control: How injected surfactants and bio stimulants may be forced into low-permeability units. Remediation 13(3):59–66
16. Kavanaugh MC, Rao PSC, Abriola L, Cherry J, Destouni G, Falta R, Major D, Mercer J, Newell C, Sale T, Shoemaker S, Siegrist RL, Teutsch G, Udell K (2003) The DNAPL cleanup challenge: source removal or long-term management. EPA/600/R-03/143, Dec 2003
17. Krembs FJ, Siegrist RL, Crimi M, Furrer RF, Petri BG (2010) *In situ* chemical oxidation for groundwater remediation: analysis of field applications and performance experiences. Ground Water Monit Remediation 30(4):42–142

18. Lake LW (1989) Enhanced oil recovery. Prentice Hall, Englewood Cliffs
19. Lee ES, Seol Y, Fang YC, Schwartz FW (2003) Destruction efficiencies and dynamics of reaction fronts associated with permanganate oxidation of trichloroethylene. Environ Sci Technol 37(11):2540–2546
20. Li XD, Schwartz FW (2000) Efficiency problems related to permanganate oxidation schemes. In: Wickramanayake GB, Gavaskar AR, Chen ASC (eds) Chemical oxidation and reactive barriers: remediation of chlorinated and recalcitrant compounds. Battelle, Columbus, pp 41–48
21. Loden P, Han Q, Porta L, Urteaga I, Barnhart K, Hakkarinen D, Illangasekare T, Jayasumana AP (2009) A wireless sensor system for validation of real-time automatic calibration of groundwater transport models. J Syst Software 82(11):1859–1868
22. Lowe KS, Gardner FG, Siegrist RL, Houk TC (2000) EPA/625/R-99/012 U.S. EPA Office of Research and Development, Washington, DC
23. McCray JE, Munakata-Marr J, Silva JAK, Davenport S, Smith MM (2010) Multi-scale experiments to evaluate mobility control methods for enhancing the sweep efficiency of injected subsurface remediation amendments. Final report for SERDP project ER-1486. Arlington. http://www.serdp.org/content/download/9275/110619/file/ER-1486-FR.pdf
24. Musaloiu-ER, Terzis A, Szlavecz K, Szalay A, Cogan J, Gray J (2006) Life under your feet: a wireless soil ecology sensor network. In: Proc. Third Workshop on Embedded Networked Sensors (EmNets)
25. Oesterreich RC, Siegrist RL (2009) Quantifying volatile organic compounds in porous media: effects of sampling method attributes, contaminant characteristics and environmental conditions. Environ Sci Technol 43(8):2891–2898
26. Porta L, Illangasekare TH, Loden P, Han Q, Jayasumana A (2009) Continuous plume monitoring using wireless sensors: Proof of concept in intermediate scale tanks. ASCE J Environ Eng 135(9):831–838
27. Ramanathan N, Balzano L, Burt M, Estrin D, Harmon T, Harvey C, Jay J, Kohler E, Rothenberg S, Srivastava M (2006) Rapid deployment with confidence: Calibration and fault detection in environmental sensor networks. Technical report for Laboratory for Embedded Collaborative Systems (LECS), University of California, Los Angeles
28. Reitsma S, Marshall M (2000) In: Wickramanayake GB, Gavaskar AR, Chen ASC (eds) Chemical oxidation and reactive barriers: remediation of chlorinated compounds. Battelle, Columbus, pp 25–32
29. Scheutz C, Broholm MM, Durant ND, Weeth EB, Jørgensen T, Dennis P, Jakobsen CS, Cox E, Chambon J, Bjerg PL (2010) Field evaluation of biological enhanced reductive dechlorination of chloroethenes in clayey till. Environ Sci Technol 44(13):5134–5141
30. Seright RS, Martin FD (1991) Fluid diversion and sweep improvement with chemical gels in oil recovery processes. Second annual report (DOE/BC/14447-10), Contract No. DE-FG22-89BC14447, U.S. DOE
31. Siegrist RL, Urynowicz MA, West OR, Crimi ML, Lowe KS (2001) Principles and practices of *in situ* chemical oxidation using permanganate. Battelle, Columbus, 336 pp
32. Siegrist RL, Crimi M, Petri B, Simpkin T, Palaia T, Krembs FJ, Munakata-Marr J, Illangasekare T, Ng G, Singletary M, Ruiz N (2010) *In Situ* chemical oxidation for groundwater remediation: Site specific engineering and technology application. Project ER-0623 Interactive CD prepared for the ESTCP, Arlington. http://serdp-estcp.org/
33. Siegrist RL, Crimi M, Simpkin TJ (eds) (2011) *In situ* chemical oxidation for groundwater remediation. Springer, New York, 678 pp
34. Silva JAK (2011) The utility of polymer amendment for enhancing *in situ* remediation effectiveness. PhD dissertation, Colorado School of Mines, Golden
35. Smith MM, Silva JAK, Munakata-Marr J, McCray JE (2008) Compatibility of polymers and chemical oxidants for enhanced groundwater remediation. Environ Sci Technol 42(24):9296–9301
36. Smith MM, Silva JAK, Munakata Marr J, McCray JE (2009) The use of polymer solution for enhanced PCE biodegradation in heterogeneous aquifer settings. *In Situ* and on-site bioremediation 10th international symposium, Battelle, May 5–8, Baltimore

37. Sorbie KS (1991) Polymer-improved oil recovery. CRC Press, Boca Raton
38. Stroo HF, Ward CH (eds) (2010) *In Situ* remediation of chlorinated solvent plumes. Springer, New York, 786 pp
39. Trubilowicz J, Cai K, Weiler M (2009) Viability of motes for hydrological measurement. Water Resour Res 45(6):16
40. Tsitonaki A, Petri B, Crimi M, Mosbaek H, Siegrist RL, Bjerg PL (2010) *In situ* chemical oxidation of contaminated soil and groundwater using persulfate: a review. Crit Rev Environ Sci Technol 40(1):55–91
41. Werner-Allen G, Lorincz K, Welsh M, Marcillo O, Johnson J, Ruiz M, Lees J (2006) Deploying a wireless sensor network, IEEE internet computing, March/April, 2006, pp 1069–7801
42. West OR, Cline SR, Holden WL, Gardner FG, Schlosser BM, Thate JE, Pickering DA, Houk TC (1998) ORNL/TM-13556, Oak Ridge National Laboratory, Oak Ridge
43. West OR, Siegrist RL, Cline SR, Gardner FG (2000) The effects of *in situ* chemical oxidation through recirculation (ISCOR) on aquifer contamination, hydrogeology, and geochemistry. Oak Ridge National Laboratory report submitted to DOE, Office of Environmental Management

Part III
National Policies

Chapter 16
Overview of U.S. EPA and Partner Information Resources Regarding Groundwater Cleanup Technologies

Walter W. Kovalick, Jr., Ph.D. and Linda Fiedler

Abstract National legislation first passed in the United States in 1980 creating programs to assess and cleanup sites contaminated from abandoned and on-going industrial processes, natural resource extraction industries, and urban sources of pollution such as gasoline stations and dry cleaners. It was followed by both revisions and elaboration at the federal level as well as similar legislation in most states in the years following. During the first years of these cleanup programs, efforts concentrated on the sites with the greatest risk, largely from direct exposure to drums, abandoned waste materials, large scale lagoons, spills, etc. Technologies for these problems were largely derived from classic civil engineering approaches (e.g., secure land disposal or engineered barriers) as well as available destruction technologies (i.e. incineration). As these obvious risks were controlled, the less visible contamination to groundwater rose in prominence, and existing remediation approaches (mainly pumping and treating) showed their limitations both in terms of cost and effectiveness.

As U.S. EPA and others were faced with reducing costs and increasing the effectiveness of remediation of these contamination problems, they began to develop and organize cost and performance information on remediation technologies on a more real time basis. With EPA regional offices, state partners, other Federal agencies, consulting engineers and parties responsible for these contamination problems as prime "clients", EPA headquarters devoted resource to developing cost and performance information on monitoring and measurement and clean up technologies to help users. This chapter summarizes the large body of information resources.

W.W. Kovalick, Jr., Ph.D. (✉)
U.S. Environmental Protection Agency, 77 W Jackson Blvd. MJ-9, Chicago, IL, USA
e-mail: kovalick.walter@epa.gov

L. Fiedler
U.S. Environmental Protection Agency, Office of Solid Waste and Emergency Response,
Technology Innovation Office, Washington, DC 20460, USA
e-mail: fiedler.linda@epa.gov

F.F. Quercia and D. Vidojevic (eds.), *Clean Soil and Safe Water*, NATO Science for Peace and Security Series C: Environmental Security, DOI 10.1007/978-94-007-2240-8_16, © Springer Science+Business Media B.V. 2012

16.1   Introduction and Background

The first national legislation in the U.S. to spur cleanup of contaminated soil and groundwater from the mismanagement of hazardous wastes and materials was passed just over 30 years ago. This statute, Comprehensive Environmental Response Compensation and Liability Act of 1980 (also known as Superfund due to the initial use of dedicated taxes for a trust fund to finance the program), launched a program at the federal level. This program enabled and partnered with similar programs at the state and tribal level over the ensuing years. Contaminated site cleanup is a collective effort among federal, state, local, and tribal governments; parties responsible for creating the sites; and the consulting and engineering communities. This combined endeavour has led to the evolution of an extensive body of practice for the assessing, monitoring, and ultimately the remediating of thousands of contaminated properties.

With the maturation of the internet in the mid-1990s, information on the scientific and engineering experiences of many parties moved from printed reports to electronic availability. In addition, dedicated efforts by government, researchers, and others to compile cost and performance information on monitoring and remediation approaches created fact sheets, documents, databases, web sites, and searchable information sources that made technical, engineering, and cost information much more broadly available.

In the context of the NATO Advanced Research Workshop on Drinking Water Protection by Integrated Management of Contaminated Land, this chapter focuses on selected U.S. EPA and partner internet resources that contain specialized information on the assessment of groundwater contamination and its subsequent control and remediation. The purpose of this chapter is to inform readers on the depth of internet resources on these topics and how to locate them on the web.

16.1.1   Groundwater Contamination

Many of the most difficult contamination challenges resulting from mismanagement of hazardous waste derive from the assessment and cleanup of groundwater. In particular, as the practice has evolved some of the most problematic contaminants are dense non-aqueous phase liquids (DNAPLs) because they are persistent in the environment and their behaviour in the subsurface is much less predictable due to their physical and chemical characteristics (e.g., denser than water, relative degrees of insolubility). Common, widely used industrial solvents, including trichloroethylene and perchloroethylene typify this class. Complicating their characteristics is the variety of subsurface regimes that range from less complicated single layer sands to complex clay and sand to fractured bedrock. Thus, some of the most challenging site cleanups result from the combination of these difficult contaminants trapped in these hard to assess subsurface regimes. While the internet resources discussed in this chapter also contain many resources related to above

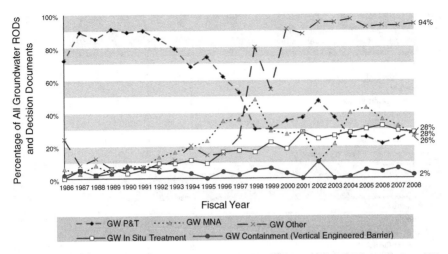

Fig. 16.1 Trends in RODs and decision documents selecting groundwater remedies (FY 1986–2008)

and below surface treatment of soils and above ground treatment of groundwater (for instance, using pump and treat), the resources specifically highlighted will be targeting information on cost and performance of those technologies that can treat DNAPLs in situ.

16.1.2   Superfund Sites

The U.S. EPA Superfund program is targeted at a priority list of sites called the National Priority List (NPL) that has evolved since 1980 using a scoring system coupled with other factors to prioritize resources. As of September 2010, U.S. EPA has listed 1,627 sites on the NPL and has brought 475 of those sites into "site wide ready for anticipated use" [2]. U.S. EPA publishes detailed information on the status of cleanups, technologies used, and other factors regularly. The most recent summary report and the figures for this chapter are taken from Superfund Remedy Report, 2010 [3].

Figure 16.1 outlines overall groundwater technology trends associated with Superfund site cleanups since 1986. This figure tracks some 1,727 decision documents. (In many cases, there is more than one decision document per Superfund site.)

Of special note is the early dependence on (followed by a tapering off of) pumping and treating groundwater at the surface. This trend is evidence that it took a number of years for the effectiveness of pump and treat to be understood and for research to yield a suite of newer technology options – especially for the more complex sites with DNAPLs. Of special note as well is the growth in the selection of in situ treatment technologies, up to 28% of total remedies, as well as groundwater – other, which represents the growth in the use of institutional controls (e.g., land use

Remedy Type and Technologies	2005	2006	2007	2008	Total
Groundwater Pump-and-Treat	22	20	23	18	83
In Situ Treatment of Groundwater	24	31	28	18	101
Bioremediation	13	20	17	12	62
Chemical Treatment	9	11	14	4	38
Air Sparging	5	2	1	2	10
Permeable Reactive Barrier	3	3	1	1	8
Phytoremediation	0	2	1	0	3
Fracturing	1	0	0	0	1
Multi-Phase Extraction	1	0	0	0	1
Unspecified Physical/Chemical Treatment	0	0	1	0	1
MNA of Groundwater	34	35	30	17	116
Groundwater Containment (Vertical Engineered Barrier)	4	4	6	1	15
Other Groundwater	73	90	88	61	312
Institutional Controls	63	79	77	52	271
Monitoring	62	80	58	39	239
Alternative Water Supply*	6	6	5	9	26
Engineering Control**	0	1	3	0	4
Total of Remedy Types	157	180	175	115	627

* Decision documents may be counted in more than one category.
* Decision documents include RODs, ROD amendments, and select ESDs.
* Alternative water supply includes alternative drinking water, well head treatment, installation of new water supply wells, increasing capacity of existing water treatment plant, and treat at use location.
** Engineering control includes sewer/sump abandonment and the use of trees for hydraulic gradient control.

Fig. 16.2 Remedy types in decision documents selecting groundwater remedies (2005–2008)

and access restrictions [often coupled with other technology approaches]) to attain protective levels at sites.

As to the types of technologies used for groundwater treatment in the Superfund program, Fig. 16.2 provides a breakdown of the various in situ, containment, and other related technologies used in the program in the most recent period from 2005 to 2008. This period was selected as representative of the more typical remedies selected in recent years vs. using the entire history of the program.

16.2 U.S. EPA and Partner Internet Resources

This section describes the location and content of four major internet sites that have extensive multi-media information resources related to the cost and performance of monitoring and cleanup of groundwater with special emphasis on DNAPLs. U.S. EPA either manages these sites or is partnered with others to consolidate information from joint data gathering efforts. There are literally hundreds of fact sheets, documents, assessments, reports, and live and archived web seminars on these sites. Of special note is the site managed by the association of states – Interstate Technology Regulatory Council – dealing with these cleanup issues; U.S. EPA is a major collaborator along with industry and researchers for materials presented on this site.

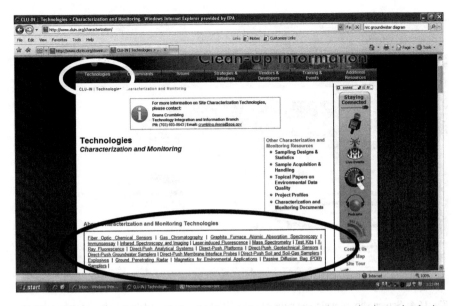

Fig. 16.3   Screen shot of clu In web site showing choices of groundwater monitoring technologies

16.2.1   U.S. EPA Clean Up Information (CLU In)

This web site (https://www.cluin.org) is one of the most extensive ones containing information on monitoring and measurement and cleanup of contaminated soil and groundwater. U.S. EPA collaborates with other entities – especially other federal and state agencies, foreign organizations, industry associations, and others to maintain a vast collection of reports, documents, fact sheets, live and archived webinars, and web links.

Over 550 pdf documents and 360 internet seminars are available for downloading and review. Over 322,000 documents and 580,000 podcasts were downloaded in the last fiscal year. Over 100 free internet seminars are delivered each year reaching more than 14,000 participants. The seminars are usually two hours in length and consist of state of the practice briefings by researchers or practitioners on the cutting edge of new monitoring or remediation approaches. In addition, EPA operates a free monthly email update to alert subscribers to new developments, such as documents, webinars, and conferences. This service called, Tech Direct, can be subscribed to at: http://www.cluin.org/techdirect/.

Of special interest for this article is the fact that there are over 2,700 resources listed in www.cluin.org based on a search for the term – groundwater. While this may include some duplicative results, there is still a very robust set of materials for those interested in monitoring and cleaning up groundwater.

Focusing on groundwater resources, Fig. 16.3 is the screen shot for the Clu In site after the "Technologies" tab in the upper left corner is selected, followed by the choice of "Characterization and Monitoring." (See http://www.clu-in.org/characterization/.)

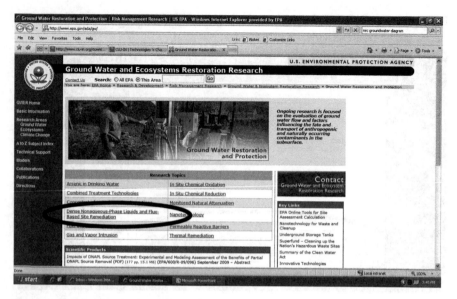

Fig. 16.4 Screen shot of welcome page of U.S. EPA Robert S. Kerr laboratory research web site

At the bottom of the screen shot are some 18 technology choices that lead to fact sheets and other information on the use of these technologies.

Similarly, by choosing Remediation under the upper left tab "Technologies", a list of soil and groundwater remediation technologies is listed including about 14 technologies for contaminated groundwater. (See http://www.clu-in.org/remediation/.)

16.2.2   U.S. EPA Groundwater and Ecosystem Research

A second web site with significant research information about groundwater characterization and remediation is operated by U.S. EPA's Office of Research and Development's Robert S. Kerr Laboratory. (See www.epa.gov/ada/gw.). This site is a compilation of current and completed research on the latest developments in groundwater monitoring and cleanup being conducted by EPA scientists. As shown in Fig. 16.4, the front page on the site offers further choices on 12 topics related to groundwater, including one on DNAPLs and flux-based site remediation. Figure 16.5 is the screen shot for this subsidiary DNAPL page, which contains further detailed references and journal articles.

16.2.3   Federal Remediation Technologies Roundtable (FRTR)

A third major resource for applied information on full scale monitoring and remediation projects is the site sponsored by the Federal Remediation Technologies Roundtable (http://www.frtr.org).

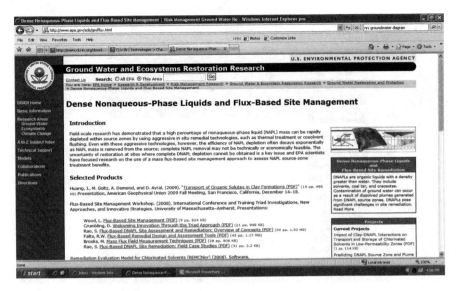

Fig. 16.5 Screen shot of DNAPL research reference page

Created in 1990, the Roundtable brings together information from the principal organizations of the U.S. Government that have responsibilities for cleaning up contaminated soil and groundwater on their own property.

In addition to EPA, the contributing members are the U.S. Air Force, Army, Navy, Departments of Energy and Interior, and the National Aeronautics and Space Administration.

The web site consists of searchable databases related to the cost and performance of both monitoring and remediation projects conducted with public funds, summary reports evaluating technologies, a collection of free models and decision support tools, a technology screening matrix to evaluate the applicability of some 64 technologies to certain problem sets, cost estimating tools as well as links to other resources. To give some sense of the depth of the resource, there are more than 400 searchable remediation and almost 200 site characterization and monitoring case studies in their respective databases on the site. In addition, there are nearly 100 "assessment" reports evaluating various remediation technologies. Also, with regard to groundwater systems, there are more than 100 reports on optimizing the performance of remediation systems.

Of particular note for this chapter are the cost and performance databases that are listed on the menu of the home page for the web site (see Fig. 16.6). Selecting that "button" for case studies leads to a choice of case study databases and selecting the FRTR Searchable Database gives the image shown in Fig. 16.7. This query allows the user to search one or more parameters at the same time including: site name, media, contaminants, primary technology, supplemental technology, the U.S. state in which the project was conducted as well as a general field for keywords in the text.

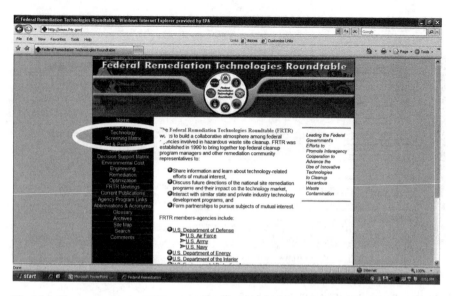

Fig. 16.6 Home page for the FRTR site with Cost and performance case studies listed on menu at the left

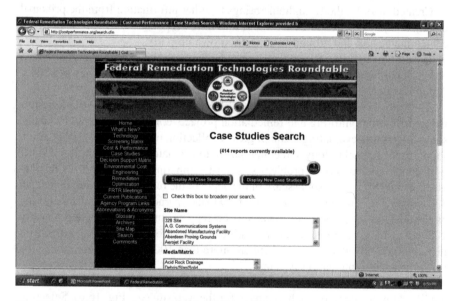

Fig. 16.7 Opening screen for search of 414 FRTR remediation case studies

A sample search using DNAPLs as the contaminant and in situ bioremediation yielded four projects shown in Fig. 16.8. Drilling down into each one of these items yields a one to page summary and the ability to download a complete report on the project.

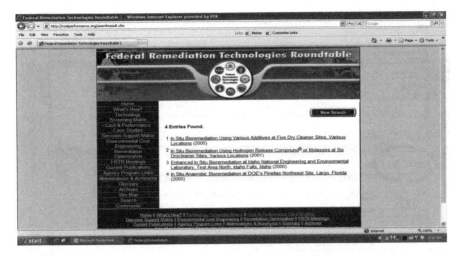

Fig. 16.8 Screen shot or search for case studies: DNAPLs using in situ bioremediation as the technologies

16.2.4   Interstate Technology and Regulatory Council (ITRC)

A fourth major web resource with a robust collection of applied information on many characterization and remediation approaches is operated by the Interstate Technology and Regulatory Council (ITRC) (see www.itrcweb.org). The ITRC is a voluntary coalition of states, federal agencies, industry and other stakeholders working to advance the application of monitoring and remediation technologies for soil and groundwater. Of special note is that the documents, training materials, and other resources are the product of consensus work among these stakeholders and are often accepted by multiple U.S. states as a baseline for their environmental decision making for projects within their jurisdiction.

Figure 16.9 is the screen shot of the welcome page for the web site; note the Guidance Document button on the left margin. Figure 16.10 is a partial screen shot of the page of guidance documents. There are some 35 documents listed on this page related to both soil and groundwater. Of these, some 17 documents relate to groundwater including remediation of both light and dense non-aqueous phase liquids. In addition to guidance documents, the ITRC main page also highlights free internet based training which is of special interest for readers not only because of future opportunities that are listed on the home page of the web site, but also because all past internet seminars have been recorded and archived.

Figure 16.11 is a screen shot of the main page related to internet based training with Internet-Based Training Archives highlighted at the bottom. Clicking that resource leads to the screen shot shown in Fig. 16.12, which is a page from the Clu In web site discussed above.

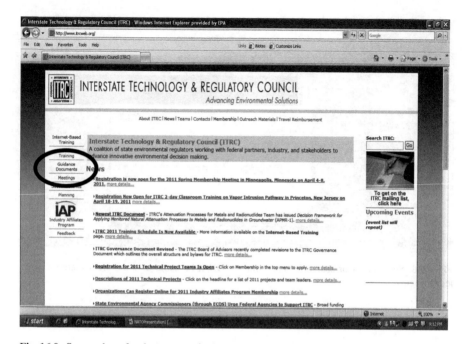

Fig. 16.9 Screen shot of welcome page for interstate technology and regulatory council

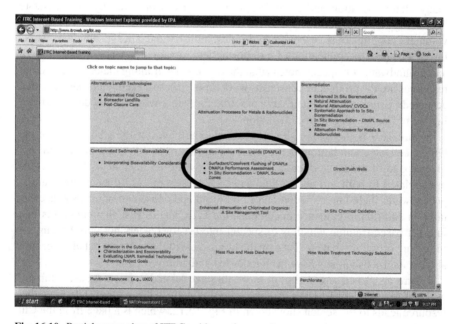

Fig. 16.10 Partial screen shot of ITRC guidance documents

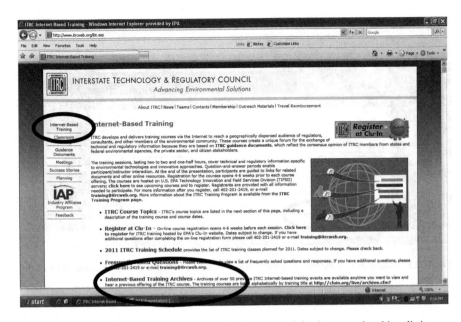

Fig. 16.11   Screen shot of ITRC web site showing internet training button and archives listing

Fig. 16.12   Screen shot of Clu In linked page with internet seminars from ITRC and others

Figure 16.13 is the screen shot of the main page for the archived internet seminars and podcasts on the Clu In web site. This page contains not only the ITRC seminars, but many more additional seminars from EPA and other sources. There are more than 350 seminars and podcasts archived on the site.

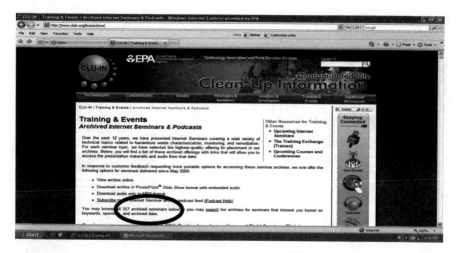

Fig. 16.13 Screen shot of the archived internet seminars and podcasts from ITRC and many other sources

16.3 Emerging Technologies and Internet References

There are several emerging and new technologies for groundwater monitoring and remediation (especially for DNAPLs), which are located on these internet resources that deserve special note.

16.3.1 Green Remediation

The maturation of engineering practice has led to efforts to improve operations and reduce energy and environmental impact of the cleanup approaches them. U.S. EPA has created a special location on its web presence of these practices under the general heading of Green Remediation. This web address is: http://www.epa.gov/superfund/green remediation/. This site contains not only EPA references, but links to efforts by standards organizations at the national and state level to better define these practices.

16.3.2 Mass Flux Measurement

With regard to the monitoring of performance of cleanups of DNAPLs and other groundwater contaminants, the classic approach of sampling wells on a site placed on a grid has turned out to be less effective because of the unique characteristics of DNAPLs and their fate and transport. A recent publication of the ITRC web site (http://www.itrcweb.org/Documents/MASSFLUX1.pdf) provides an extensive reference (over 150 pages) on the theory and practice of this approach [1].

16.3.3   Nanotechnology and Remediation

Nanomaterials are pervasive in the economy and have become a valuable tool for dealing with contaminated groundwater. U.S. EPA has created a separate page in Clu In on applications of nanotechnologies to environmental remediation. (See http://www.clu-in.org/techfocus/default.focus/sec/Nanotechnology:_Applications_ for_Environmental_Remediation/cat/Application/#2).

The table of contents on this page includes the following subject areas: Nanoscale Materials, Contaminants Known to be Treated Using Nanoscale Materials, Reactive Chemistry of Nanoscale Materials, In Situ Application of Nanoscale Materials, Factors Affecting Performance, Fate, Transport, and Toxicity, Field Demonstrations and Case Studies, Cost, Nanotechnology Vendors, and Nanotechnology Products with Potential Remediation Applications.

16.4   Summary

Researchers, engineering practitioners, project managers, private sector firms, and government officials needing up to date and applied information on monitoring and measurement as well as remediation of contaminated soil and groundwater can find a wealth of resources outside the published literature. Efforts by the public sector and others in the U.S. have led to the accumulation of hundreds of fact sheets, documents, web pages, databases, internet seminars (both on-going and archived), and podcasts that provide state of the practice updates.

16.5   Endnote

The opinions expressed in this chapter are those of the authors and do not necessarily represent the views of the U.S. Environmental Protection Agency.

References

1. Interstate technology and regulatory council (2010) Use and measurement of mass flux and mass discharge, Aug 2010 (http://www.itrcweb.org/Documents/MASSFLUX1.pdf)
2. U.S. EPA (2010) Office of solid waste and emergency response fiscal year 2010 end of year report, 2010. (http://www.epa.gov/oswer/docs/oswer_eoy_2010.pdf)
3. U.S.EPA (2010) Superfund remedies report. 13th editon EPA542-R-10-004, Sept. 2010 (http://www.cluin.org/asr)

Chapter 17
Towards Sustainable Contaminated Sites Management in Austria

Harald Kasamas, Gernot Döberl, and Dietmar Müller

Abstract Since more than 20 years Austria has a specific legal and funding framework for contaminated sites management. In 2008 the Ministry of Environment took the initiative to revise the legal framework and prepare a new policy based on the experiences made and taking into account international developments in industrial countries over the past years. The main goal is to proceed with the current contaminated land programme towards an integrated system of contaminated land management supporting environmentally friendly approaches and enhancing cost-efficiency.

17.1 Introduction

The Act on the Remediation of Contaminated Sites (ALSAG) came into force in 1989 [1]. ALSAG regulates the identification and assessment of suspected contaminated sites and the remediation of high-priority contaminated sites in Austria. The main purpose of ALSAG is to establish a financing system to ensure that urgent measures at the most risky sites can be taken as quickly as possible. In addition, ALSAG implements a country-wide uniform methodology for the identification, assessment and prioritisation of these sites.

ALSAG regulates the management of contaminated sites where contamination took place before 1989. So, ALSAG refers to the management of "historical contamination". New pollution is dealt with in other environmental laws.

H. Kasamas (✉)
Division of Contaminated Land Management, Ministry of Environment,
Stubenbastei, A-1010 Vienna, Austria
e-mail: harald.kasamas@lebensministerium.at

G. Döberl • D. Müller
Contaminated Sites Department, Environment Agency, Spittelauer Lände 5,
A-1090, Vienna, Austria

F.F. Quercia and D. Vidojevic (eds.), *Clean Soil and Safe Water*, NATO Science for Peace and Security Series C: Environmental Security, DOI 10.1007/978-94-007-2240-8_17,
© Springer Science+Business Media B.V. 2012

Apart from defining the general framework for Contaminated Sites Management, the Act primarily provides the legal basis to obtain grants for funding remediation activities.

17.2   The Financial Model

Funds for remediation projects are raised essentially by levies for land filling, waste treatment and waste export. The created funds are earmarked for cleanup operations and assessment activities to identify high-priority contaminated sites. The funds are collected and distributed at the federal level. This ensures a country-wide uniform procedure.

The total revenue for contaminated sites management amounts to 1,055 billion Euro or 1.5 billion US Dollar (1990–2011). The annual average is around 67 mio Euro or 95 mio US Dollar. The yearly distribution is shown with Fig. 17.1.

It can be seen that the annual amount is changing over the years. The reason is that the collection of waste levy is incentive driven. This means, higher levies have to be paid where waste treatment practice is not meeting State-of-the-Art levels. So one incentive is to raise technical standards at waste treatment facilities; another one is to minimise land filling. These incentives were quite successful, which is good news for waste management standards, but less for obtaining the needed funds for contaminated sites remediation.

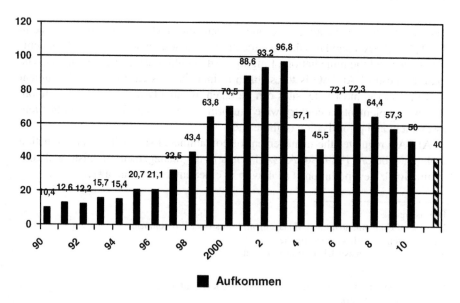

Fig. 17.1 Annual income for the Austrian cleanup act (1990–2011)

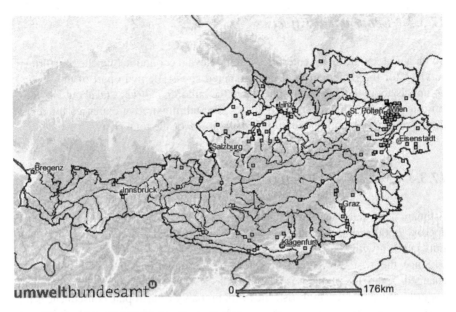

Fig. 17.2 Location of high-priority sites in Austria

17.3   Review of Achieved Results [6 – 8]

17.3.1   Contaminated Sites Management

Over the past 20 years of Contaminated Sites Management, 58,000 suspected sites have been identified through systematic investigations; 500 risk assessments performed; 255 high-priority sites identified and 200 high-priority sites remediated (see Fig. 17.2).

17.3.2   Environmental Effects

- The remediation of contaminated sites enabled a qualitative restoration of groundwater to the extent of about 46 million m³. This amount of water corresponds to the annual consumption of about 1.7 million people.
- 9.9 million tons contaminated soil were excavated and safely disposed.
- Through the remediation of contaminated sites the exhaust of greenhouse gases, especially methane, has been prevented. These measures contributed to a reduction of annual greenhouse gas emissions by 0.3% in the years 2004 and 2005.
- 145 ha of Brownfield land has been redeveloped and returned into the economic cycle. These measures reduced Greenfield consumption and further sealing.

17.3.3   Economical Effects

So far about 350 different enterprises from relevant economic branches performed services for the remediation of contaminated sites. The main part of the service volume is about 490 mio. Euro allotted to the building industry and about 380 mio. Euro to the waste management industry. The administrative expenses for registration, evaluation and remediation are about 3% of the total costs.

17.3.4   Future Challenges

Estimation by the Austrian Environment Protection Agency predicts a total of approx. 2,500 seriously contaminated sites in Austria and additionally 5–6 billion Euros to deal with those.

Since the past 20 years, perception of the contaminated sites problem has changed. Today the problem is recognised as a widespread structural problem and not just limited to a few high-risk sites with probably catastrophic consequences for humans and environment in earlier times. In this regard, a huge amount of sites has been predicted which requires a lot of resources. Consequently, the management process has to speed up and sufficient financial means have to be ensured [5].

It has been recognised that contaminated sites have also considerable consequences for spatial planning beside environmental threats. The economic loss of abandoned and underused properties due to possible contamination is substantial. Stimulation of Brownfield redevelopment is needed and would have also an environmental benefit for soil protection by reducing sealing of Greenfields [10].

17.3.5   Legal Needs

By history, the Austrian Water Act enacted in 1959 forms the main legal basis for contaminated sites in Austria. Its main goal is to protect water resources and enforces a 'zero'-pollution tolerance for groundwater ("has to be kept in natural state"). Nowadays, the Waste Management Act – enacted in 1990 – and subsequent ordinances are another legal driver to protect public interests and to reduce environmental burdens.

However, both laws are based on precautionary principles with the aim to prevent contamination. But already contaminated soil and groundwater are a result where precautionary principle has failed. Therefore, to set remediation targets based on a precautionary principle is in many cases neither technically nor economically possible.

This is the legal background to revise the existing contaminated land policy in Austria and set guidelines which allow for risk-based remediation targets [9].

17.4 Towards Sustainable Site Management

17.4.1 Mission Statements

In 2008, a policy/science working group developed a new vision based on sustainable principles for contaminated sites management in Austria [3]. Various stakeholders have been included in this process for commends and discussions of the interim and final results. In 2009, the Environment Ministry published the final product as a policy paper [2]. Six Mission Statements have been developed in order to guide towards sustainable management of contaminated sites in Austria. These six statements are:

- The inventory of historically contaminated sites shall be completed within one generation (intended timeframe 2025).
- Measures (decontamination, containment, monitoring, and use restrictions) at seriously contaminated sites shall be designed within two generations (timeframe 2050).
- Risk assessments have to be realized by site and use specific conditions.
- Measures shall allow site specific and land use related conditions into account. Risks for human health or the environment must be adequately managed.
- Remediation measures (decontamination, containment) need to be sustainable with lasting effects to enhance the environmental status of a site.
- Framing conditions for reusing and integrating contaminated sites back into economic cycles shall be improved.

The cited key objectives are the basis for the legislative drafting process on a major revision of the current law (ALSAG). It will encompass risk-based and sustainable contaminated site management solutions. The amended law is expected to get in effect in 2013.

17.4.2 Technical Basis

Beside the political discussion the preparation of scientific and technical guidance documents for practical implementation has already started in 2008. In this regard, a major project [4] has been launched which involves key scientists and experts to work on the key areas of risk-based land management [11]. These key areas and the developed guidelines are listed in Fig. 17.3.

The products of the project are:

- Guidelines for human health risk assessment
- Risk-based screening values for soil
- Report on Ecological Risk Assessment
- Guidelines for assessing leaching conditions in the unsaturated zone

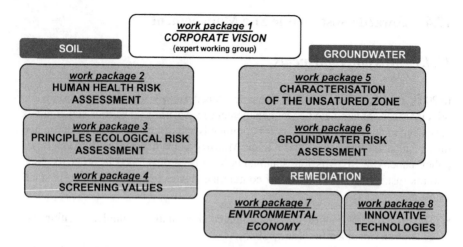

Fig. 17.3 Project structure of Altlastenmanagement 2010

- Guidelines for groundwater risk assessment
- Decision-Support-Tool for evaluating and selection of remediation options (cost-benefit analysis)
- Platform on sustainable remediation technologies (Workshops, Bulletins, RTD Recommendations)

The interim and final results are presented and discussed in joint workshops including public authorities (both on federal and provincial level), industrial representatives, scientists and practitioners to obtain an integrated view and to involve all experts who are responsible for setting the new concept into practice.

17.5 Prospect

It is expected that the new contaminated sites policy will become operative in 2013. The draft text of the law refers to two technical ordinances, which will build on the outcome of the project Altlastenmanagement 2010. It is expected that revised policy will lead to a more sustainable, environmentally and economical sound decision making.

References

1. Bundesgesetz vom 7 Juni 1989 zur Finanzierung und Durchführung der Altlastensanierung (Altlastensanierungsgesetz). BGBl. Nr. 299/1989 i. d. g. F (1989)
2. Bund-Länder-Arbeitskreis, Managementkontaminierter Standorte: Leitbild Altlastenmanagement. Bundesministerium für Land- und Forstwirtschaft, Umwelt und Wasserwirtschaft (Hrsg)

3. Döberl G, Mueller D, Achleitner G, Kastner, M. Enna, Kasamas H, Punesch J, Repetschnigg W (2008) A new model for the contaminated land management in Austria . Contaminated spectrum issue 1 / 2009, ITVA, Erich Schmidt Verlag

4. Doberl G, Muller D, Kasamas H (2010) Austria in transition. CONSOIL 2010 Proceedings, Salzburg

5. Ferguson C, Darmendrail D, Freier K, Jensen BK, Jensen J, Kasamas H, Urzulai A, Vegter JJ (eds) (1998) Risk assessment for contaminated land in Europe, vol 1, Scientific basis. LQM Press, Nottingham

6. Kommunalkredit Public Consulting & Umweltbundesamt (2007) Altlastensanierung in Österreich – Effekte und Ausblick. Bundesministerium für Land- und Forstwirtschaft, Umwelt und Wasserwirtschaft (Hrsg.), Wien

7. Müller D, Prokop G, Weihs S (2009) Policy and legal framework for contaminated sites in Austria – achievements so far and future directions. Proceedings of the International Conference on Contaminated Sites, Bratislava

8. Prokop G, Skala Ch, Kasamas H (2008) Austria draws balance after 19 years contaminated sites financing law. Altlastenspektrum Ausgabe 2/2008. ITVA. Erich Schmidt Verlag

9. Reichenauer TG, Wimmer B, Rachor I, Friesl-Hanl W & Loibner AP (2007) Nutzungsbezogene Bewertung von Altlasten und kontaminierten Standorten – Handlungsempfehlungen für die Ableitung standortspezifischer Sanierungsziele unter Berücksichtigung der derzeitigen bzw. geplanten Nutzung. Umweltbundesamt (Hrsg)

10. Umweltbundesamt (2008) (Bau) Land in Sicht. Gute Gründe für die Verwertung industrieller und gewerblicher Brachflächen. Bundesministerium für Land- und Forstwirtschaft, Umwelt und Wasserwirtschaft (Hrsg), Wien

11. Vegter JJ, Lowe J, Kasamas H (eds) (2003) Sustainable Management of Contaminated Land: An Overview. Austrian Federal Environment Agency on behalf of CLARINET, Vienna

Chapter 18
Assessment of Sites Under Risk for Soil Contamination in Serbia

Dragana Vidojević

Abstract The database of potentially polluted and polluted sites in Serbia contains geo-referenced and non geo-referenced data and information about potentially contaminated and contaminated sites, remediation activities and the data set of subjects that caused the pollution. Soil data are collected by different institutions and from different projects. The greatest number of registered sources of localized soil pollution is related to municipal waste disposal sites, oil extraction and storage sites, industrial and commercial sites. Within industrial pollution sites, oil industry has the greatest share. Remediation goals of contaminated sites are set at the national level related.

18.1 Introduction

The Serbian Environmental Protection Agency (SEPA) is a governmental institution under the supervision of the Ministry of Environment, Mining and Spatial Planning. Its main tasks are: development, management and coordination of the national environmental information system, collection and integration of environmental data, production of annual state of the environment reports and providing recommendations for future steps aiming at a general improvement in this field.

Since 2006 the Serbian Environmental Protection Agency started the creation of a database on potentially polluted and polluted sites as a part of the national Environmental Information System [1]. At the moment, the database covers the industrial localities that were identified until 2010 and waste disposal sites identified from 2009 to present.

D. Vidojević (✉)
Environmental Protection Agency, Ministry of Environment, Mining and Spatial Planning, Ruže Jovanovića 27a, Belgrade 11160, Serbia
e-mail: dragana.vidojevic@sepa.gov.rs

F.F. Quercia and D. Vidojevic (eds.), *Clean Soil and Safe Water*, NATO Science for Peace and Security Series C: Environmental Security, DOI 10.1007/978-94-007-2240-8_18, © Springer Science+Business Media B.V. 2012

The development of the database is in accordance with requirements of the Guidelines for European Environment Information and Observation Network (EIONET) data collection on contaminated sites. The database consists of:

– Sources of contamination (recognized and potential): municipal and industrial waste disposal sites, industrial and commercial areas, oil extraction and storage sites, oil spills, power plants and mining sites.
– Status of recognized contaminated sites: existence of a preliminary study, investigation results and level of implemented remediation.

18.2 Progress Quantification in Management of Contaminated Sites

In the territory of the Republic of Serbia 375 contaminated sites have been identified [15]. From the analysis of contaminated sites management data, it can be concluded that preliminary studies on all identified contaminated sites up to 2010 have been carried out, while main site investigations have been completed on a lesser number of sites (Table 18.1). Data sources on progress in the management of local soil pollution are represented by research and monitoring projects on soil and groundwater pollution from localized sources. These data are then used as input to the database managed by the Environmental Protection Agency.

The largest number of registered sources of local soil pollution is related to municipal waste disposal sites (Table 18.2 and Fig. 18.1).

The database of municipal waste disposal sites was updated in 2009. In the Republic of Serbia, there are 164 landfills that are being used by municipal public

Table 18.1 Progress quantification in the management of local soil pollution in Serbia

Reference year	2010	
Processing step	Number of sites	
Site identification/preliminary study	Already identified	375
Preliminary investigation	Already completed	211
Main site investigation	Already completed	13
Measures completed	Already completed	12

Table 18.2 Soil polluting activities from localized sources in Serbia

Number of registered sources							
Municipal waste disposal sites	Industrial waste disposal sites	Industrial and commercial sites	Mining sites	Oil extraction and storage sites	Oil spills sites	Power plants	Total
164	31	61	5	99	2	13	375

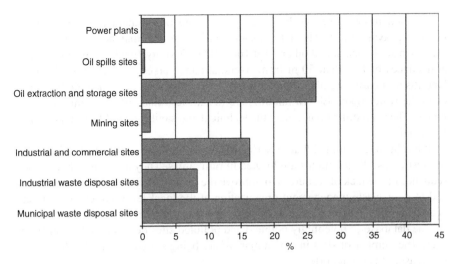

Fig. 18.1 Soil polluting activities from localized sources as % of total sites where (preliminary or main) site investigation has been completed

Fig. 18.2 Public utility companies landfill sites in Serbia

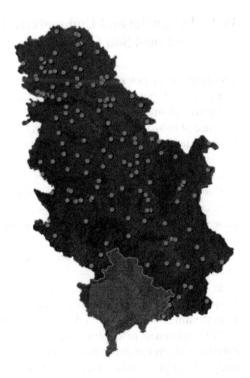

utility companies for waste disposal which represent a potential soil and groundwater pollution source. There is no detailed soil and groundwater contamination surveys on locations where municipal waste disposal sites are identified (Fig. 18.2).

Detailed information on the location, size and other characteristics of these landfills is available. Twelve landfills are located within a distance of 100 m from the urban settlements, and other 50 at 100–500 m. Twenty-five landfills are located at distances of less than 50 m from rivers, streams, lakes or reservoirs. Fourteen landfills of the total number of landfills are practically right on a water course bank or in the water body. Eleven landfills are distant less than 500 m from the water intake. These landfills do not meet the technical and sanitary conditions reported in EU standards.

It has been determined that in Serbia there are 3,251 locations of illegal dumps. In most cases, illegal dumps are located in rural areas. They are primarily the consequence of the lack of resources to improve the quality of waste collection systems and of poor waste management organization at the local level. Exact location has been determined for all illegal dump sites by the use of GPS devices.

The database of potentially polluted and polluted sites does not include military sites. The number of sites in which manure are being stored is not reliable and for that reason it is not reported.

18.3   Industrial and Commercial Branches Responsible for Local Soil Contamination

The database of potentially contaminated industrial areas was updated in 2010. There are 211 potential contaminated industrial sites in the territory of Serbia. Reported potentially contaminated sites are identified on the basis of long term presence of industrial activities and from laboratory analysis of soil and groundwater in the near vicinity of local pollution sources. Most identified polluted soil sites within the industry belong to the oil industry (59.2%), followed by the chemical industry (15.2%) and the metal industry (13.3%) (Fig. 18.3).

Complex ecological studies at oil industry production facilities in Serbia have been conducted in 2010 with the objective to assess the environmental status of these industrial facilities and the potential impacts on the environment. The studies aimed at identifying existing and potential environmental risks as well as corrective actions for the improvement of the environmental conditions at each site (Naftna Industrija Srbije (NIS) [2]).

In order to assess soil contamination, 97 samples were taken during 2010 from 12 oil industry facilities and the results have been compared with the thresholds from the limits given in the Serbian regulation [7] (Fig. 18.4).

With reference to organic parameters, concentrations of TPH $C<12$ (Total Petroleum Hydrocarbons) above reference limits have been detected in 9.28% of the analyzed samples while concentration above reference limits of TPH $C>12$ have been detected in 12.37% of the analyzed samples.

With reference to benzene, 23.71% of samples showed concentrations above the reference limits.

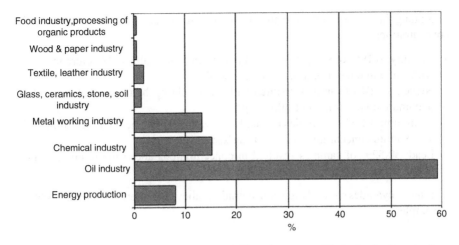

Fig. 18.3 Breakdown of industrial and commercial branches responsible for local soil contamination as % of total

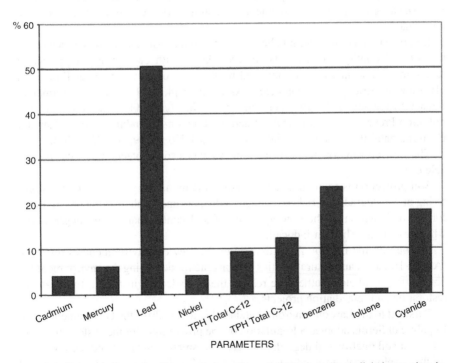

Fig. 18.4 Concentrations of pollutants above reference limits in soil at oil industry sites in Serbia

Among inorganic contaminants, the following metals were detected in high concentrations:

- Mercury: 6.19% of samples resulted in concentrations above the reference limit with a maximum concentration of 22.1 mg/kg;
- Nickel: 4.12% of samples resulted with values above the reference limit with a maximum concentration of 806.9 mg/kg;
- Cadmium: 4.12% of samples resulted with values above the reference limit with a maximum concentration of 137.1 mg/kg; and
- Lead: 50.52% of samples resulted in concentrations above the reference limit with a maximum concentration of 1,361 mg/kg.

Finally, cyanides showed high concentrations above the reference limit in 18.56% of samples.

18.4 Legal Framework for the Management of Contaminated Sites

There is no special law on soil protection in the Republic of Serbia and the administrative capacities for the enforcement of existing laws and by-laws in this area are inadequate.

Soil management is accomplished by planning the sustainable use and conservation of soil quality and diversity, in accordance with environmental protection requirements and measures established by the Law on Environmental Protection [4], Law on Strategic Environmental Assessment [4, 7], Law on Environmental Impact Assessment ([4]; Official Gazette of RS, No. 36/09, 2009), Law on Integrated Pollution Prevention and Control [4] and the Law on Amendments to the Law on Environmental Protection (Official Gazette of RS, No. 36/09, Art. 22, 2009).

The Law on Environmental Protection regulates soil protection and its sustainable use.

Soil protection is accomplished by systematic monitoring of soil quality, monitoring of soil degradation risk assessment indicators and implementing programmes for the remediation of the consequences of soil contamination and degradation, whether natural or human-induced.

Pursuant to the Law on Amendments to the Law on Environmental Protection (Article 16), legal and natural entities responsible for degrading the environment are obliged to remediate or otherwise restore the degraded environment, by means of restoration and remediation projects.

Pursuant to the Law on Environmental Protection, in 2010 the Government of the Republic of Serbia adopted a Regulation on the programme for the systematic monitoring of soil quality, soil degradation risk assessment indicators and methodology for the development of remediation programmes [7]. The Regulation is harmonized with the recommendations given in the Proposal for a European Soil Framework Directive – COM (2006) 232.

The adoption of this Regulation has provided the tools to ensure soil protection -based on prevention of degradation - through identification of soil degradation risk areas, whether such degradation is natural or human-induced. The Regulation provides the basis for identification and management of contaminated sites in the territory of the Republic of Serbia. The level of chemical contamination of soil is assessed on the basis of limit values of contaminant concentration listed in the Regulation's Annex. In order to design contaminated soil remediation projects, additional studies to assess the level of soil contamination at the identified contaminated sites are carried out. According to the Regulation, the inventory of contaminated sites is an integral part of the environmental protection information system administered by the Environmental Protection Agency. The development of large scale soil surveys together with the Soil information System should support the decision making and land management system in Serbia [10].

Restoration and remediation priorities are identified on the basis of the Regulation which establishes criteria for the assessment of the status of highly threatened environment, the status of threatened environment and establishes criteria for the identification of restoration and remediation priorities (Official Gazette of RS, No. 22/10, 2010). The National Sustainable Development Strategy of the Republic of Serbia [5], the National Environmental Protection Program [9] and the Spatial Development Strategy of the Republic of Serbia [6], represent the strategic background for soil protection in the Republic of Serbia.

The strategic objectives of sustainable land use are given in the National Sustainable Development Strategy of the Republic of Serbia [5]. The objectives include:

1. Harmonization of legislative acts related to land use and soil protection with EU legislation;
2. Prevention of further soil loss, especially due to industrial, mining, energy, communication and other activities, and conservation and enhancement of soil quality;
3. Prevention of soil degradation, changes of land use and development and management of agricultural soil.

In the framework of goals set up in the National Environmental Protection Program of the Republic of Serbia (2010), the goals related to the management of contaminated sites include rehabilitation of trash dumps - which are the greatest risk for environment - remediation of contaminated soils in industrial complexes and remediation and recultivation of areas degraded by mining activities.

For the waste management planning, the Waste Management Strategy for the Period 2010–2019 was adopted in May 2010 [8]. This norm represents the basic document that provides conditions for rational and sustainable waste management in the Republic of Serbia, estimates the progress in waste management, lays down short-term and long-term objectives and provides conditions for rational and sustainable waste management.

The general objective in the Waste Management Strategy is the development of a sustainable waste management system with the aim to reduce environment pollution and land degradation. One particular (short-term) objective (2010–2014) is to construct 12 regional waste management centres until 2014 (regional landfills, plants for separation of recyclable waste, plants for biological waste treatment and transfer stations in each region).

18.5 Conclusion

According to the data on inventory of contaminated sites collected by the Environmental Protection Agency, it can be concluded that:

1. From available data, a list of contaminated sites has been created. At these sites analyses of soil and groundwater quality have confirmed an increased level of pollution.
2. However the data collected so far are not at the same level of quality at different locations, therefore it is not possible to estimate a comparable level of pollution at different contaminated sites, nor the impact on environmental media and human health.
3. In the past, due to the lack of reference thresholds the list of contaminated sites was created on the basis of different criteria.
4. Additional and more detailed surveys are needed today in order to update the inventory of all sites.
5. The new legislation enacted in 2010 established the definition of contaminated sites together with reference values and provided a legal background for future prioritization studies and detailed investigations.
6. Resulting from these surveys, a National priority list for restoration and remediation of most polluted localities will be created.

References

1. Dedijer A, Mitrovic-Josipovic M, Radulovic E, Dimic B et al (2007) Environment in Serbia – an indicator – based review. Ministry of Science and Environmental Protection/Environmental Protection Agency, Belgrade. ISBN 978-86-84163-34-1
2. Naftna Industrija Srbije (NIS) (2010) Complex ecological examination of Naftna Industrija Srbije (NIS) production facilities, Novi Sad, Serbia
3. OGRS (1994) Book of rules about the allowed amounts of dangerous and harmful substances in soil and water for irrigation and methods for their testing, vol 23/94. Official Gazette of the Republic of Serbia, Belgrade
4. OGRS (2004) Law on environmental protection, law on strategic environmental assessment, law on environmental impact assessment and law on integrated pollution prevention and control of the environment, vol 135/04. Official Gazette of the Republic of Serbia, Belgrade
5. OGRS (2008) National sustainable development strategy for the period 2009–2017, vol 57/08. Official Gazette of the Republic of Serbia, Belgrade
6. OGRS (2008) Spatial development strategy of the Republic of Serbia, vol 119/08. Official Gazette of the Republic of Serbia, Belgrade

7. OGRS (2010) Regulation on the programme for the systematic monitoring of soil quality, soil degradation risk assessment indicators and methodology for the development of remediation programmes, vol 88/10. Official Gazette of the Republic of Serbia, Belgrade
8. OGRS (2010) Waste management strategy for the period 2010–2019, vol 29/10. Official Gazette of the Republic of Serbia, Belgrade
9. OGRS (2010) National environmental protection programme, vol 12/10. Official Gazette of the Republic of Serbia, Belgrade
10. Protic N, Martinovic Lj, Milicic B, Stevanovic D, Mojasevic M, (2005) The status of soil surveys in Serbia and Montenegro. In: Jones RJA, Houšková B, Bullock P, Montanarella L (eds) Soil resources of Europe, 2nd edn. European Soil Bureau Research report no. 9, EUR 20559 EN. Office for Official Publications of the European Communities, Luxembourg, pp 297–315
11. Report on the state of the environment of the Republic of Serbia in 2006, 2007, 2008 and 2009, Ministry of Environment, Mining and Spatial Planning, Environmental Protection Agency, Republic of Serbia, Belgrade, http://www.sepa.gov.rs/
12. Vidojević D (2006) Report on Progress in management of contaminated sites CSI 015 indicator. Ministry of Science and Environmental Protection-Environmental Protection Agency, Republic of Serbia, Belgrade
13. Vidojević D, Manojlovic M, (2007) Overview of soil information and policies in Serbia, Chapter in the book: Status and prospect of soil information in south-eastern Europe: soil databases, projects and applications, Institute for Environment and Sustainability, EC JRC, EUR 22656 EN
14. Vidojević D (2005) Soil information system in Serbia. XI Congress of Soil Protection, Book of abstracts, Budva
15. Vidojević D (2009) Report on the state of soil of the Republic of Serbia. Ministry of Environment and Spatial Planning, Environmental Protection Agency, Belgrade. ISBN 978-86-87159-02-0

Chapter 19
Geochemistry of Bottled Waters of Serbia

Tanja Petrović, Milena Zlokolica Mandić, Nebojša Veljković,
Petar Papić, and Jana Stojković

Abstract Chemical analyses of 13 bottled mineral waters were carried out at the BGR geochemical laboratories. The analyses included pH, electrical conductivity, alkalinity and concentrations of 69 elements and ions. An aquifer lithology impacts on the chemical composition of ground water significantly, especially on the explanation of conditions of forming and circulation of ground water through different lithology environments. Basic composition of ground water is usually a reflection of the lithogeochemistry of the aquifer, while micro components indicate the circulation of ground water through the different lithological environment. The waters are most frequently tapped from Neogene carbonate rocks (dolomite, limestone), and to a lesser extent from granitoid rocks, shale, and serpentinite. Based on the analyses of bottled mineral waters, it has been observed that water quality is greatly affected by the chemical composition of igneous intrusions, regardless of the fact that the analysed waters have been sampled from different aquifers (Neogene sediments, limestone, flysch, schist). Bottled waters of Serbia are mostly HCO_3-Ca, HCO_3-Ca-Mg (from carbonate rocks) and HCO_3-Na (from Neogene and igneous rocks). Among the micro components, increased concentrations of Cs, Ge, Rb, Li, and F are frequently present in bottled water, as a consequence of its circulation through granitoid rocks. Some samples contain a higher concentration of B, I, NH_4, Tl, W, as the consequence of the aquifer environment.

T. Petrović (✉) • M.Z. Mandić
Geological Institute of Serbia, Rovinjska 12, Belgrade 11000, Serbia
e-mail: tanjapetrovic.hg@gmail.com

N. Veljković
Serbian Environmental Protection Agency, Ministry of Environment Protection,
Ruže Jovanovića 27a, Belgrade 11160, Serbia

P. Papić • J. Stojković
Faculty of Mining and Geology, University of Belgrade, Đušina 7, Belgrade, Serbia

F.F. Quercia and D. Vidojevic (eds.), *Clean Soil and Safe Water*, NATO Science for Peace   247
and Security Series C: Environmental Security, DOI 10.1007/978-94-007-2240-8_19,
© Springer Science+Business Media B.V. 2012

19.1 Introduction

Previous investigations of the chemical composition, genesis and geochemistry of mineral waters in the region of Serbia were described by [16, 17, 29, 38–40], [8, 26, 27, 30, 31, 36], and [43].

There was not any specific review on bottled mineral water within the stated investigations until the completion of the Geochemistry of European Bottled Water Project [32] by the European Geological Surveys (EGS) Geochemistry Expert Group. The final project data set included 884 bottled waters from 40 European countries, among which 13 bottled mineral waters were from the territory of Serbia (Fig. 19.1). The first paper on the 'Hydrogeological Conditions for the Forming and Quality of Mineral Waters in Serbia' was published by Petrović et al. [28].

Currently, there are 30 mineral water bottling plants, which delivered about 635 million litres of bottled water to the market in 2010. Looking back, the production of natural mineral water in the year 2000 was 330.3 million litres, and 539 million litres in 2005 [44]. It is quite obvious that the consumption of bottled water is increasing rapidly [33]. This unprecedented rise is largely due to advertising, and to the poor quality of tap water. On the Serbian market, there is a great variety of bottled waters, where carbonated acid waters hold 50% of the retail market. Additionally, the proportion of bottled waters with high mineralization is significant and preferred in Serbia, unlike low mineralized bottled waters that prevail in Western European markets [32].

19.2 Geology of Serbia

The territory of Serbia is divided into five regional geotectonic units (Fig. 19.1): the Pannonian Basin, the Interior Dinarides, the Vardar zone, the Serbian–Macedonian massif and the Carpatho-Balkanides, which are significantly different according to the quantity and quality of ground water.

Bottled waters in Serbia are abstracted from a diverse range of aquifers (Fig. 19.1). The most significant aquifer is the Mesozoic limestone, which mainly occurs in the eastern (Carpatho-Balkanides) and western parts (Interior Dinarides) of Serbia.

In the Carpatho-Balkanides, Mesozoic limestone and dolomite of more than 1,000 m thickness are widely distributed.

In the Interior Dinarides, there is a significant occurrence of Triassic limestone and dolomite, followed by the Jurassic diabase-chert formation (ophiolitic belt) with subordinate limestone in the overlying parts, and the Cretaceous formations with predominately flysch [42].

The Vardar Zone borders the Serbian Massif and the Dinarides. It is characterized by a complex structure consisting of medium grade metamorphosed schist of yet undetermined age. Recrystallized limestone and marble are also present within the schist [42].

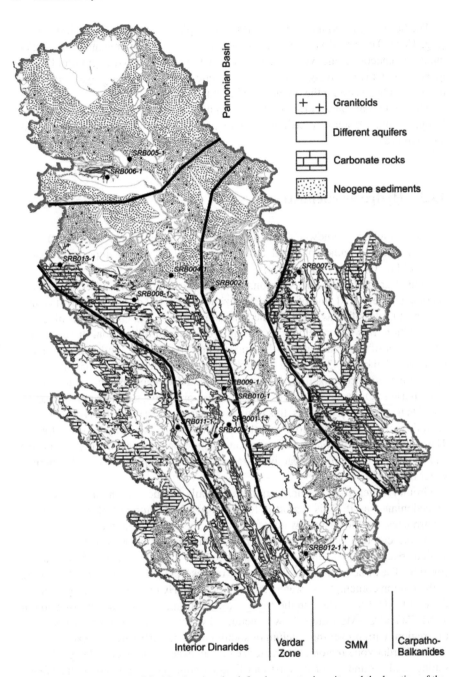

Fig. 19.1 Geological map of Serbia showing the defined geotectonic units and the location of the 13 bottled waters

The Serbian-Macedonian Massif (SMM) occupies the central part of the territory (Fig. 19.1). The SMM is composed of very thick Proterozoic metamorphic rocks: gneiss, micaceous shale, various types of schist, marble, quartzite, granitoid rocks, igneous rocks (intrusive-granitoid and volcanic rocks of Tertiary age), etc. Deep reverse faults constitute the boundary with other geotectonic units.

The Pannonian Basin, or its south-eastern part in Serbia, consists of Palaeogene, Neogene and Quaternary sediments with a total maximum thickness of about 4,000 m [18].

19.3 Hydrogeochemistry

Low-mineralized groundwater is mainly used for a water supply (tap water) or as mineral water in balneology and for bottling. Ground water temperature ranges from cold (5°C) to thermal (111°C).

In the Carpatho-Balkanides region, ground water from karst aquifers prevails. It is comparatively cold, and of low mineralization (TDS 0.2 do 0.4 g/L); by its anion content it is classified as HCO_3, while dominant cation is Ca. On the basis of its gas composition, this water belongs to the nitrogen type. In this region, there are also occurrences of thermal water with temperatures above 20°C, and with the maximum recorded being 43°C. Their mineralization ranges from 0.3 to 0.65 g/L. On the basis of the anion composition, this water is of the HCO_3 type, and more rarely it has an increased SO_4 ion concentration, as a result of, primarily, sulphide mineral oxidation. On the basis of its cations, this water is classified as Ca or Na-Ca type. Nitrogen prevails in the gas composition, with the frequent occurrence of hydrogen-sulphide. The occurrence of elevated concentrations of microcomponents, such as fluorine and boron, is common. Radioactivity (radon) is comparatively high in the thermal water of karst terrains.

Ground water from fissured aquifers prevails in the region of the Serbian-Macedonian massif. This region is famous for a large number of mineral water occurrences. On the basis of water temperature and gas content, these waters range from cold carbonic acid mineral waters to high thermal nitrogen waters (111°C). With respect to the pH value, the waters range from highly acid (pH 3.5) to alkaline (pH 9.3). The mineralisation of these waters ranges from 0.2 to over 5 g/L. According to their anion content, these waters are mainly of the HCO_3 type, and more rarely of the HCO_3-SO_4 type, while on the basis of their cation content, there are singled out as Na-Ca and Ca-Mg waters. Low mineralized waters (lesser than 0.5 g/L) occur in this region, with prevailing calcium or sodium ions, but there are also exceptions.

The Vardar zone region is characterized by an exceptionally complex geological setting, and ground water in karst and fissured aquifers is quite common. Waters originating from limestone are of the Ca, Ca–Mg or Na type, while their anion content is mainly HCO_3. The diversity of cations originates from the hydrogeological conditions of circulation and interaction with different rocks (igneous, metamorphic, sedimentary), which occur within the watershed area, and more or less participate in the aquifer formations. There are more than 80 mineral water occurrences in fissured aquifers, comprising igneous rocks ranging from alkaline to felsic compositions, but also a whole range of metamorphic and sedimentary rocks. On the basis of the overall mineralisation, these waters range from low mineralised (0.3 g/L) to highly mineralised (5.2 g/L). In the Vardar zone, there are numerous occurrences of thermal waters reaching a maximum temperature of 68°C. HCO_3 ions prevail in the anion composition, but HCO_3–SO_4 and HCO_3–Cl waters occur as well. Na, Na-Ca and Na-Mg waters characterize the cation composition. Waters are predominantly of carbonic acid, but there are occurrences of a nitrogen composition. The occurrences with an extremely high content of hydrogen-sulphide should be pointed out.

Within the region of Inner Dinarides, there are karst aquifers. Springs are characterised by a high yield (several hundreds L/s). Ground waters are mostly of low mineralization (0.2-0.3 g/L), and the temperatures range from 5 to 36°C. Exceptionally, there are waters with a temperature of 37°C and mineralization up to 0.5 g/L. These are of HCO_3-Ca type, and more rarely of the HCO_3-Ca-Mg type.

Within the Pannonian basin is ground water of high mineralization in shallow parts of structures with mineralization up to 1 g/L. This is conditioned by slower processes of water replenishment. Within the Neogene basin, there occur frequent alternations of sediments with variable filtration characteristics. Accordingly, waters are of the HCO_3-Na or HCO_3-Cl–Na type. In the deeper parts of the basin, ground waters change their chemical properties from Cl-HCO_3-Na, to Cl-Na and Cl-Ca-Na types. In still deeper parts, the ground water is of a brine type and increased temperatures. Nitrogen, nitrogen-methane, and methane waters occur as well. The content of organic matters is significantly higher in relation to other regions due to sedimentation conditions.

19.4   Sampling Collection and Analytical Methods

During 2008, 13 bottles of selected mineral waters of Serbia were sent to the laboratory of the Federal Institute for Geosciences and Natural Resources (BGR) in Berlin, Germany, where they were analysed [1, 32]; 12 samples in 0.5 l PET bottles were bought from markets, and one was collected from its spring (Iva voda) before the beginning of the commercial sale period.

The water bottling localities, as well as the primary aquifer from which the water was extracted, are presented in Table 19.1.

Seventy-two parameters were determined on the bottled water samples. The following techniques were used for the analysis of bottled waters: inductively coupled plasma quadrupole-mass spectrometry (ICP-QMS, trace elements), inductively coupled plasma atomic emission spectroscopy (ICP-AES, major elements), ion chromatography (IC, anions), atomic fluorescence spectrometry (AFS, for Hg), titration (alkalinity), photometric methods (NH4), potentiometric methods (pH) and conductometric methods (EC). Detailed analytical procedures are described by [1, 32].

Table 19.1 Analysed bottle waters of Serbia (*Gruss, the fragmental products of in-situ granular disintegration of rocks, usually granite)

Sample	Name	Bottling location	Aquifer geology	Well/spring	Type
SRB001-1	Voda Kopaonik	Graševačka reka-Brus	Granitoid	Well	Non-carbonated
SRB002-1	Aqua Balkanika	Cerovac-Smederevska Palanka	Neogene sediments	Well	Non-carbonated
SRB003-1	Eva	Brzeće-Brus	Limestone breccia in the lower part of santonian (K_2^3) flysch deposits	Well	Non-carbonated
SRB004-1	Dar voda	Darosava-Aranđelovac	Neogene sand-clay deposits, above granite-monzonite	Well	Natural CO_2
SRB005-1	Minaqua	Novi sad	Neogene sediments	Well	Non-carbonated
SRB006-1	Jazak	Jazak-Fruška gora	Triassic limestone, dolomite	Well	Non-carbonated
SRB007-1	Duboka	Neresnica-kučevo	Limestone	Well	Natural CO_2
SRB008-1	Voda voda	Gornja Toplica-mionica	Triassic limestone	Well	Non-carbonated
SRB009-1	Voda vrnjci	Vrnjci-kraljevo	Palaeozoic schist and amphibolite, with layers of marble, serpentinite	Well	Natural CO_2
SRB010-1	Mivela	Trstenik	Serpentinite, schist	Well	Natural CO_2
SRB011-1	Iva	Gornji gradac-Raška	Phyllite, schist and metasandstone above middle Triassic limestone	Spring	Non-carbonated
SRB012-1	Bivoda	Rakovac-Bujanovac	Granitoid gruss* above Hercynian granitoid	Well	Natural CO_2
SRB013-1	Tronoša	Korenita	Triassic limestone, dolomite	Spring	Non-carbonated

19.5 Regulations, Standards and Law

19.5.1 Regulations on Ground Water Quality

There are two regulations in Serbia concerning the quality of water for human usage. These are: Regulation on the Hygienic Acceptability of Potable Water [22, 23], and Regulation on Quality and Other Requirements for Natural Mineral Water, Spring Water and Bottled Drinking Water [24]. The comparison of Regulations and Standards in Serbia with those of EU Directives and World Health Organization [41] is shown in Table 19.2.

The Regulation on the Hygienic Acceptability of Potable Water defines the Maximum Admissible Concentrations (MAC) of chemical substances in water for public water supply, and in addition, it provides MAC for certain metals in bottled water (Al, Ba, Ca, Cl, CN, Cu, F, Fe, Hg, Ng, Mn, Na, Ni, N_2, NO_3, SO_4, Zn, and electrical conductivity less than 500 μS/cm). Except for As, B and U, all MAC values in this regulation are lower than the ones given by the World Health Organization (WHO). The Regulation on Quality and Other Requirements for Natural Mineral Water, Spring Water and Bottled Drinking Water defines MAC values for specific chemical parameters that may pose risk to human health, indicators of water quality and the nomenclature of mineral waters. If the concentration of a certain parameter in mineral water is higher than the one given in Table 19.2, (value n*), then this must be highlighted in the water name (on the bottle label). Example: "bicarbonate water" if HCO_3 = 600 mg/L, or "Mg water" if Mg = 50*, etc.

The MAC values for F (1.5 mg/L), Cl (250 mg/L), CN (70 μg/L) and SO_4 (250 mg/L) must not be exceeded, while for water rich in CO_2, the pH value can be less than 6.8. This regulation applies to all ground water, regardless of the overall mineralisation. The term "spring water" was used to indicate "captured water at the location".

The regulation is harmonised with WHO standards, except in the case of the B content, for which a maximum concentration is allowed that is two times higher.

19.5.2 Regulations for Opening and Monitoring Springs in Exploitation

According to the laws and regulations of the Republic of Serbia all ground water being exploited are subject to the submission of a ground water reserve feasibility study as the first of a set of documents to acquire the right for exploitation. The reserves of ground water imply an average minimal ground water quantity per month, expressed in L/s of which at least 90% can be obtained from one watershed, a water-bearing environment, a singled out deposit or part of a deposit during the lowest water level, thus the exploitation does not result in quality deterioration [21].

Table 19.2 Comparison of regulations and standards in Serbia with those of EU Directives and WHO

Parameter	Unit	EU directive 1998/83/EC DRINKING WATER	EU Directive 2003/40/EC MINERAL WATER	EU Directive 2009/54/EC NATURAL MINERAL WATER	WHO	Measurements concentration in analytic samples Min	Measurements concentration in analytic samples Max	Regulation on the hygienic acceptability of potable water [22, 23] Mac of chemical substances in water for public water supply	Regulation on the hygienic acceptability of potable water [22, 23] Mac for certain minerals in bottled water	Regulation on quality and other requirements for natural mineral water, spring water and bottled drinking water (Official Gazette of Serbia and Montenegro, volume 53/05)[24]
pH		6.5–9.5	n.d.	-		5.6	7.5	6.8–8.5	6.8–8.5	6.8–8.5
EC at 20°C	mS/cm	2500 g.v.	n.d.	-		340	4,560	<1,000	<500	2,500
Ag	mg/l	n.d.	n.d.	-		<0.001	0.00357		10	
Al	mg/l	200 g.v.	n.d.	-	200	<0.001	8.38	200	50	200
As	mg/l	10	10	-	10	0.085	6.26	10	50	10
B	mg/l	1000	-	-	500	22.3	5,660	300	1,000	1,000
Ba	mg/l	n.d.	1	-	0.7	<0.01	0.637	0.7	0.1	
Be	mg/l	-	-	-		<0.001	0.5012		0.2	
Br	mg/l	-	-	-		-	-			
Ca	mg/l					22.2	241	200	100	150*
Cd	mg/l					<0.001	0.00589	3	5	3
Cl	mg/l					1.35	287	200	25	200*–250
CN-	mg/l	50	70	-		-	-	50	-	50–70
Cr	mg/l	50	50	-	50	0.09	2.4	50	50	50
Cs	mg/l	n.d.	n.d.	-		0.0425	84.7			
Cu	mg/l	2000	1000	-	2000	0.113	2.48	2,000	100	2,000
F	mg/l	1.5	5(1.5: label)	<1	1.5	<0.002	2.39	1.2	1	1*–1.5
Fe	mg/l	200 g.v.	n.d.	-	300	0.13	101	300	50	200
Fe²⁺	mg/l	n.d.	n.d.	<1		<0.002	2.39			
Ge	mg/l	n.d.	n.d.	-		0.00532	17.8			1*
HCO₃⁻	mg/l	n.d.	n.d.	<600		200	3,290			600*
Hg	mg/l	1	1	-	6	<5	<5	1		1

Element	Unit									
I	mg/l	n.d.	n.d.	-		1.68	686			
K	mg/l	n.d.	n.d.	-		0.8	52	12	10	
Li	mg/l	n.d.	n.d.	-		0.762	985			
Mg	mg/l	n.d.	50 g.v.	<50	50	12.08	324	50	30	50*
Mn	mg/l	500	n.d.	-	400	<1	465	50	20	50
Mo	mg/l	n.d.	200 g.v.	<200	70	0.131	0.537	70		
Na	mg/l	n.d.	0.5 g.v.	-	200	1.8	1,216	150	20	200*
NH_4^+	mg/l	n.d.	20 g.v.	-		<0.005	4.4			0.5
Ni	mg/l	20	0.5 g.v.	-	70	0.0246	9.12	20	10	20
NO_2^+	mg/l	0,1	50 g.v.	-	3	<0.005	0.144	0.03	-	0.1
NO_3^+	mg/l	50	10	-	50	<0.01	9.25	50	5	50
Pb	mg/l	10	n.d.	-	10	0.00297	0.0855	10		10
Rb	mg/l	n.d.	n.d.	-		0.388	205			
Si	mg/l	n.d.	5	-		(SiO_2) 4.8	(SiO_2) 88.8			
Sb	mg/l	5	n.d.	-	20	0.161	2.93	3	10	5
Sc	mg/l	n.d.	10	-		0.0194	0.348			
Se	mg/l	10	250 g.v.	-	10	<0.01	0.149	10	10	10
SO_4	mg/l	n.d.	n.d.	<200	250	0.02	173		25	200*–250
Sr	mg/l	n.d.	n.d.	-		0.05	1.49			
Tl	mg/l	n.d.	n.d.	-		0.00441	1.11			
U	mg/l	n.d.	n.d.	-	15	0.00166	1.83		50	
V	mg/l	n.d.	n.d.	-		0.037	4.45		1	
W	mg/l	n.d.	n.d.	-		0.0182	11			
Zn	mg/l	n.d.	n.d.	-	3000	0.0609	3.27	3.000	100	

Note: n.d. – not detectable; g.v.-groundwater; n*- low boundary nomenclature
MAC- Maximum acceptable concentrations

While working on the feasibility study, it is necessary to examine ground water quality at three month intervals during one hydrological year (13 to 15 months), as well as to monitor the regime (the quantity and ground water table every five days). In addition, it is necessary to define the hydrogeological conditions (recharging, circulation, drainage), and the hydrogeological parameter of the environment, the relationship with surface water and other relevant parameters. The feasibility study on ground water reserves is completed every five years and defended in front of the State Commission within the Ministry in charge (in the past 10 years, depending on the organization of the government of Serbia, the competence for ground water exploitation was within different ministries).

For the protection measures of water supply springs, the Regulation on Way of Determination and Maintenance of Sanitary Protection of Water Supply Springs Zone [25] is applied. This Regulation covers both springs, which due to their quantity and quality can be used for public drinking water, and bottling water springs. In order to protect water, three sanitary protection zones are established, depending on specific hydrogeological conditions (karst, a free water table aquifer, an aquifer under pressure). This document describes compulsory measures and activities that must be carried out or prohibited in some zones, as well as the way of zone marking on the terrain. The feasibility study on spring sanitary protection zones is defended in front of the Commission of the Ministry of Health of the Republic of Serbia only after estimating the reserves. The local government is in charge of carrying out of determined measures and spring sanitary protection zones.

In addition to the above stated conditions, it is necessary to provide the opinion and approval of the following institutions and ministries in charge of culture, wildlife protection, city planning, civil engineering, waterpower engineering and health (usability and quality control of products for consumption).

Although the regulations required to obtain a water-bottling license are strict, the problems arise after, due to the absence of coordination and competence in application, and the lack of rigorous penalties for endangering the springs.

19.6   Results of Chemical Analyses

19.6.1   PH, EC and Macrocomponents

Values of PH, EC and macrocomponents in ground water are given in Table 19.3.

The PH value of the analysed water ranges from 5.6 to 7.8, with a median of 6.93. The low PH value is due to the presence of CO_2 (Dar voda, Minaqua, Voda Vrnjci, Mivela, and Bivoda).

The TDS value ranges from 202.6 to 3400.8 mg/L, with a median of 924.13 mg/L. Generally, bottled waters of Serbia are enriched in dissolved solids, as the consequence of aquifer lithology and residence time of ground water. Waters with high TDS usually occur in metamorphic and igneous environments, while waters with low TDS occur in carbonate rocks.

Table 19.3 Results of pH, EC, TDS and macro component of the analysed samples

Sample	Name of waters	PH	EC (μS/cm)	TDS (mg/L)	Ca (mg/L)	Mg (mg/L)	Na (mg/L)	K (mg/L)	Cl (mg/L)	SO$_4$ (mg/L)	HCO$_3$ (mg/L)
SRB001-1	Voda kopaonik	7.5	1,700	1115.5	28.3	12.8	409	7.4	18.9	0.32	1,183
SRB002-1	Aqua balkanika	7.6	672	394.8	78.5	28.3	33.9	2.1	5.2	18	440
SRB003-1	Eva	7.8	340	202.6	47.8	15.2	3.1	1	1.8	15.8	200
SRB004-1	Dar voda	5.6	990	637.5	90.6	22.6	92.4	17.2	28.4	80.5	521
SRB005-1	Minaqua	5.75	1,974	1181.7	22.2	19.9	412	3.6	287	0.35	768
SRB006-1	Jazak	7.5	690	410.6	77	45.6	6.9	3.5	5.5	29.5	427
SRB007-1	Duboka	6.9	1,365	864.9	241	19.7	55	5.1	15.4	9.61	956
SRB008-1	Voda voda	7.5	623	384.9	78.3	14.8	41.5	3.1	7.5	13	392
SRB009-1	Voda vrnjci	6.4	1,696	1174.8	76.3	55.4	241	35.1	15.5	29.1	1,177
SRB010-1	Mivela	6.3	2,510	1620.3	26.3	324	120	8.4	12.8	0.02	2,047
SRB011-1	Iva	7.3	423	260.5	58.7	20.7	3.4	0.8	1.97	7.26	275
SRB012-1	Bivoda	6.5	4,560	3400.8	85.4	20.6	1,216	52	54.1	173	3,290
SRB013-1	Tronoša	7.45	630	364.6	83	38.7	1.8	0.6	1.35	22.2	401
Median	Serbia	6.93	1397.9	924.1	76.4	49.1	202.7	10.7	35	30.6	929
Median	Europe[a]	6.8	588		65.9	16.4	15.5	2.1	13.4	20	

TDS = Σ macro components + Σ microelements + trace elements

[a]From publication 'Geochemistry of European Bottled Water' [32]

The reaction between ground water and aquifer minerals affects ground water quality significantly, but is also useful to understand the genesis of ground water [3] Macro components of ground water are controlled by weathering of rocks (water-rock interactions). Thus, prevailing components in water show the major impact of aquifer lithology. If the median values of some macro components of bottled waters of Europe are compared with those in bottled waters of Serbia, it is obvious that Serbian waters are richer in Mg, Na, and K, besides high values of EC and TDS.

The Piper diagram (Fig. 19.2) shows that almost all waters are of the bicarbonate (HCO_3) type, except one sample which is HCO_3, Cl type. The content of cations is wide; the prevailing cations are Na and K ions in waters from granitoids, Ca and Ca-Mg in water from carbonate environments, and Mg in only one sample (Mivela) from serpentinite rocks. Na is dominant in mineral waters (>1 g/L), while cation Ca and Ca-Mg in low-mineralised waters (<1 g/L).

The genetic diagram [6], based on macro components, has proved to be very reliable for the better understanding of aquifer lithology (Fig. 19.3), especially for occurrences formed in more complex conditions [19]. Figure 19.3a shows the shapes of lines for waters extracted from granitoid lithologies. Whereas, Fig. 19.3c shows the lines for carbonate rocks, and it is obvious that the water Duboka (SRB007-1) has a similar shaped line to those of Fig. 19.3a, which indicates the circulation of water through granitoid.

Analyzing the proportion of rCa/rMg (meq/L) in waters extracted from carbonate lithologies, a distinction can be made between those derived from limestone with rCa/rMg > 1.5 (Eva, Iva, Voda Voda), and the ones from dolomite and dolomitic limestone with rCa/rMg < 1.5 (Jazak, Tronoša).

Figure 19.3b shows circulation of water through complex geological conditions, whereas Fig. 19.3d displays the situation of waters captured from Neogene sediments, but their chemical composition mainly depends on rocks below the Neogene formations.

19.6.2 Specific Components and Trace Elements

A few elements, such as Cs, Ge, I, Li, K, Rb, Sr and Si (Table 19.4) are more widely distributed in some Serbian bottled waters than in other European bottled waters. Further, in some bottled waters, the concentration of B, Mn and NH_4 is higher than the limits allowed by Serbian regulations and the WHO.

Higher concentrations of Cs, Li, Rb, Ge and Sr indicate that the water source is in felsic rocks, mostly in granitoid intrusions (Beaucaire and Michard 1982). The main source of these elements is the weathering of K-feldspat and mica. The increased content of Li, Rb and Cs is associated with CO_2–Na waters [14]. Higher values of these elements mainly occur in ground water of Central Serbia (Voda Kopaonik, Dar voda, Voda Vrnjci, Mivela, Bivoda), where water is taken from deep fractured hard rocks. All waters (except Voda Kopaonik) are rich in CO_2 and are highly mineralised.

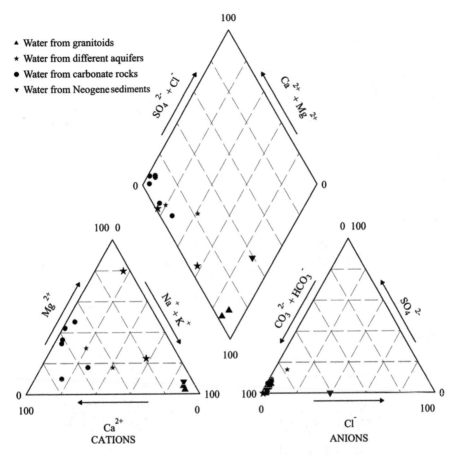

Fig. 19.2 Piper diagram of Serbian bottled waters according to the type of aquifer (Large symbol marks water with TDS>1 g/L, while small symbol marks water with TDS<1 g/L)

Fluorine is an element that is also found in this type of water. Fluorine is ascribed to granitoid minerals, such as apatite, biotite, fluorite [4, 8, 27]. Fluorine concentrations vary from 0.005 to 2.39 mg/L. For health reasons, the recommended value of F in water should range between 0.7 and 1.5 mg/L. This is the optimum value for the development of bones and teeth. Two bottled waters in Serbia, however, maintain values of over 1.5 mg/L. Small children should not drink these waters on daily basis [20] in order to prevent dental fluorosis problems [2].

The highest boron value is detected in Bivoda, and slightly lower values are found in Voda Kopaonik and Minaqua. According to one theory, the high proportion of boron indicates that the mineralisation is derived from the magnesite complex of Neogene sediments (in case of Bivoda and Minaqua), since high concentrations of boron and the boron mineral searlesite [$NaBSi_2O_5(OH)_2$] occur

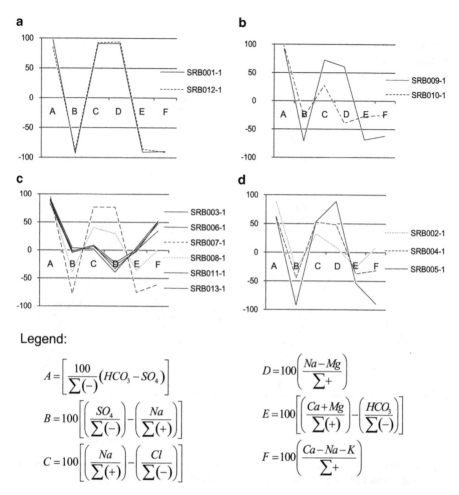

Fig. 19.3 Genetic diagram based on cations and anions expressed in %eq. (according to [6]) (**a**) Samples captured from granitoids; (**b**) Samples captured from different aquifers; (**c**) Samples captured from carbonate rocks; (**d**) Samples captured from Neogene sediments

[9]. According to another theory, boron in ground water is derived from tourmaline [14], which is found in pegmatite (Bivoda and Kopaonik). Bivoda is captured in the Bujanovac Basin which is rich in mineral waters, and all waters have a higher boron value. The Bujanovac Basin is comprised of granitoid layers, of alluvial and Neogene sediments, the basement of which is made from Hercynian granitoid. Pegmatite occurs in dykes, equivalent of granitoid, and has a wide occurrence in the Bujanovac Basin.

According to the Serbian Regulations [22, 23] the maximum allowed concentration for vanadium in bottled water is 1 μg/L, while in other Directives and

Table 19.4 Results of some specific elements and trace elements of the analysed Serbian bottled water samples

Samples	As (µg/L)	B (µg/L)	Cs (µg/L)	Cr (µg/L)	Ge (µg/L)	Fe (µg/L)	F (µg/L)	I (µg/L)	Li (µg/L)
Voda kopaonik	0.326	1,400	20.9	0.304	17.8	3.03	2.39	11.6	283
Aqua balkanika	0.085	92.9	0.0425	1.27	0.0068	0.277	0.005	13.6	31.2
Eva	6.26	22.3	0.596	0.967	0.0137	0.34	0.084	2.97	2.02
Dar voda	0.616	249	84.7	0.125	1.36	101	1.39	58.1	331
Minaqua	0.106	1,250	0.78	0.0905	0.22	3.29	0.459	686	72.8
Jazak	1.28	27.2	1.52	2.4	0.018	0.237	0.135	3.52	3.96
Duboka	2.28	158	3.33	0.284	0.38	1.27	0.477	10.2	43.5
Voda voda	5.33	408	38.4	0.97	1.99	0.193	0.747	4.85	200
Voda vrnjci	2.64	631	52.7	0.101	15.3	0.324	1.7	7.04	464
Mivela	0.14	783	25.9	0.152	2.65	6.85	0.155	12.8	300
Iva	2.8	39.8	0.325	1.47	0.0567	0.912	0.106	2.7	1.9
Bivoda	1.13	5,660	0.385	0.296	16.2	3.88	1.16	19.1	985
Tron voda	1.82	22.1	0.219	0.302	0.0053	0.132	0.089	1.68	0.762

Samples	Mn (mg/L)	Ni (µg/L)	Rb (µg/L)	Sb (µg/L)	Tl (µg/L)	V (µg/L)	W (µg/L)	NH$_4$ (mg/L)	SiO$_2$ (mg/L)
Voda kopaonik	0.003	0.679	18	0.359	0.005	0.868	0.071	-0.01	23.8
Aqua balkanika	-0.001	0.359	1.43	0.397	0.004	0.0371	0.018	0.006	4.8
Eva	-0.001	4.85	1.23	2.93	0.009	0.115	0.028	-0.01	9.5
Dar voda	0.465	3.68	148	0.313	0.013	0.396	0.319	0.205	23.6
Minaqua	0.025	0.0469	3.68	0.394	0.013	0.213	0.457	4.4	26.7
Jazak	-0.001	0.537	1.34	0.402	0.017	0.293	0.064	-0.01	12.1
Duboka	-0.001	0.839	17.1	0.585	0.052	1.18	0.049	-0.01	22.7
Voda voda	-0.001	2.48	17.3	1.61	1.11	0.207	11	-0.01	17.0
Voda vrnjci	0.003	9.12	205	0.41	0.011	0.085	0.231	0.012	80.4
Mivela	0.039	7.87	40.7	0.161	0.006	0.0961	0.233	0.114	64.4
Iva	0.001	1.61	0.875	0.311	0.01	0.622	0.068	-0.01	15.8
Bivoda	0.036	1.29	163	0.813	0.007	4.45	0.102	1.16	88.8
Tron voda	0.001	0.0551	0.388	0.607	0.054	0.275	0.033	-0.01	6.7

Regulations no maximum limit is quoted. Acid water from thermal springs may contain a few hundred µg/L [15]. The content of V in Bivoda is 4.45 µg/L. Although V is considered as a trace element, it is relatively abundant, even though it only rarely forms independent minerals in igneous rocks [34]. There is no data regarding V toxicity and available evidence does not indicate V as a problem in drinking water [5].

The content of NO_2 in Bivoda is 0.144 mg/L. According to [10], the European Commission established the potable water according to the following relationship $[NO_3-]/50 + [NO_2-]/3 \leq 1$. The result of this equation in SRB012-1 is 0.0572, which indicates that there are no negative effects on human health.

Silica (SiO_2) is connected with the temperature of water. High SiO_2 values are recorded in thermal water, such as Bivoda (88 mg/L) with a temperature of 43°C, and Voda Vrnjci and Mivela with a water temperature of about 20°C. Silicon gives stability to bones, elasticity to connective tissue, tautness to the skin, and it also has good influence on hair, fingernails and toenails [7].

The presence of Fe (0.101 mg/L) and Mn (0.465 mg/L) in Dar voda may be related to the weathering of granite rocks, or the presence of basic rocks which are not found on the surface, or in the borehole.

The higher content of Ni (9.12 µg/L) in Voda Vrnjci and Mivela indicates the presence of ultramafic rocks (gabbro and diabase).

Voda Voda is captured from limestone occurring at the bottom of Neogene sediments. Although the water was formed in limestone, according to the basic chemical composition: HCO_3-Ca, it is obvious that pegmatite affected its chemical composition, which is shown by the presence of Tl and W in water, as well as in the genetic diagram (Fig. 19.3c). Pegmatite does outcrop in the wider area.

Water Minaqua has been captured from the Neogene sediments of the Pannonian Basin, which explains the anion composition of HCO_3-Cl. Elevated contents of NH_4 and I have been observed in this water. Since the water is formed in an-aerobic conditions, higher concentrations of I (0.686 mg/L) and NH_4 (4.4 mg/L), which are of a natural organogenic origin have been observed in the mineral water. The higher concentrations of NH_4 indicate that the water has been formed in areas where oil and gas occurs [28]. Ammonium has a toxic effect on human health only if the intake becomes higher than the capacity to detoxify, which is about 33.7 mg of ammonium ion per kg of body weight per day [5] Chloride (Cl) indicates that the water is formed in a shallow environment of the Pannonian Basin. From the medical point of view, iodine improves memory, mood, and normal functioning of the thyroid gland, quality of hair, skin, teeth and nails, which gives this water a special medicinal value.

Compared to other analyzed waters, the Jazak water has a slightly higher content of Cr (2.4 µg/L), which is due to the gabbro that occurs east of Jazak village. The concentration of Cr in Jazak gabbro reaches 498 mg/kg [35] Water Jazak emerges through a fissured-karst aquifer of considerable yield, and according to these features the location is not typical of the Pannonian Basin.

19.6.3 Comparison of Analytical Results of This Study with Those on Bottle Labels

Analyses of bottled waters were performed most frequently in Serbian laboratories (the Institute of Serbia for Public Health Doctor Milan Jovanović Batut, the City Institute for Health Protection, Belgrade, SP Laboratory AD Bečej, the Institute for Public Health, Vranje), while some waters were analysed at the Fresenius laboratory in Germany.

Only macro components are presented on bottle labels, while fluorine, dry residues, or pH values are stated on some bottles. It is interesting that the sulphate content is not presented on most bottle labels.

Comparing the analytical results produced at the BGR laboratory (this study) with those on bottle labels (Table 19.5) it can be observed that aberrations amount to 3.2%. The lowest aberration of measured values is for HCO_3, while the highest ones are for Cl. After taking laboratory conditions into consideration, the correspondence between the results of this study and those on the bottle labels is quite good.

19.7 Concluding Remarks

The analyses of bottled mineral waters proved the direct dependence between the hydrochemical composition of source waters and the complex geological properties in which the formation and movement of waters have occurred throughout their geological history. However, neither the velocities of turnover, nor the related age of water, were analyzed in this study.

Ground water in Serbia is extracted from various geological environments in order to be bottled. HCO_3-Ca or HCO_3-Ca-Mg type waters from carbonate rocks

Table 19.5 Comparison of macro components from the BGR laboratory (this study) and the ones displayed on bottle labels

Analytical data	Parameter	Min. mg/L	Max. mg/L	Median mg/L	% difference
This study	Na	1.8	1,216	202.7	2.96
Bottle label	Na	1.8	1,160	208.7	
This study	K	0.6	52	10.7	2.71
Bottle label	K	0.6	54	10.99	
This study	Ca	22.2	241	76.4	2.38
Bottle label	Ca	23.2	236	74.58	
This study	Mg	12.8	324	49.1	1.32
Bottle label	Mg	13	330	49.75	
This study	HCO_3	200	3,290	929	0.26
Bottle label	HCO_3	220	3,233	931.5	
This study	Cl	1.3	287	35	3.2
Bottle label	Cl	1.9	283	36.12	

and HCO$_3$-Na waters from Neogene sediments, granitoid rocks and crystalline complexes are most widely distributed on the market.

The presence of serpentinite conditioned the formation of waters rich in magnesium. Bottled waters rich in magnesium are: Mivela (HCO$_3$-Mg) and Voda Vrnjci (HCO$_3$-Na-Mg).

The values of micro components and trace elements (obtained from the BGR laboratories – this study) have contributed significantly to the study of water genesis, namely the formation and circulation of ground water, and water-rock interactions.

Due to the data of this study, a completely different view on the genesis of some waters has been established. For example: Voda Voda originates from limestone found at the bottom of Neogene sediments (HCO$_3$-Ca type of water). However, in the genetic diagram, aberrations from waters typical of karst are obvious. The presence of Tl and W confirms these aberrations, and suggests that the water is in contact with pegmatite, which is observed at the surface.

Water Voda Vrnjci was formed within serpentinite, which in part, increased Mg content to water, while Palaeozoic schist and amphibolite impart increased Na content to water. The presence of the geochemical association of Cs, Rb, Si, Ge and K suggests that granitoid plays an important role in ground water formation. There is granitoid occurring on the surface south-east of the spring and well.

It is obvious that previous geological and hydro geological surveys were not able to study the genesis of ground water, because of the lack of adequate chemical analyses. This multi-parameter study has shed light on ground water genesis and water-rock interactions.

Continuation of this study is certainly required to confirm the genesis of ground water and isotope analyses.

Acknowledgments We would like to express our sincere gratitude to the EuroGeoSurveys Geochemistry Expert Group for the invitation to participate in the Project European Groundwater Geochemistry: Bottled Water. We would also like to especially thank Clemens Raimann, the leader of the EuroGeoSurveys Geochemistry Expert Group and Manfred Birke, and all the laboratory staff at the Federal Institute for Geosciences and Natural Resources in Berlin for their analytical work. We additionally thank Alecos Demetriades for his comments on the first draft of the manuscript.

References

1. Birke M, Reimann C, Demetriades A, Rauch U, Lorenz H, Harazim B, and Glatte W (2010) Determination of major and trace elements in European bottled mineral water-analytical methods. In: Birke M, Demetriades A, and De Vivo B (Guest eds), Mineral Waters of Europe. Special Issue, J Geochem Exploration 107(3):217–226
2. Bjorvatn K, Bardsen A, Thorkildsen AH, Sand K (1994) Fluorid i norsk grunn-vann en ukjent helsefaktor. (Fluoride in Nonvegian drinking water an un-known health factor in Nonvegian). Vann 2:120–128
3. Cederstorm DJ (1946) Genesis of groundwater in the coastally plain of Virginia. Environ Geology 41:218–245

4. Chae GT, Yun ST, Mayer B, Kim KH, Kim SY, Kwon JS, Kim K, Koh YK (2007) Fluorine geochemistry in bedrock groundwater of South Korea. Sci of Total Environ 385(1–3):272–283
5. Cicchella D, Albanese S, De Vivo B, Dinelli E, Giaccio L, Lim A, and Valera P (2010) Trace elements and ions in Italian bottles mineral waters: identification of anomalous values and human health related effects. In: Birke M Demetriades A, and B De Vivo (Guest eds) Mineral Waters of Europe. Special Issue, J Geochem Exploration 107(3):336–349
6. D'Amore F, Scandiffio G, Panichi C (1983) Some observations on the chemical classification of ground waters. Geothermics 12(2/3):141–148
7. DHaese PC, Lambert LV, De Broe M (2004) Silicon. In: Merian E, Anke M, Ihnat M, Stoeppler M (eds) Elements and their compounds in the environment, 2nd edn. Wiley-VCH Verlag GmbH & Co, Weinheim, pp 1273–1284
8. Dangić A, Protić D (1995) Geochemistry of mineral and thermal waters of Serbia: content and distribution of fluorine. Radovi Geoinstituta, Knjiga 31:315–323 (Serbian, with English summary)
9. Dangić A, Rakočević P (1993) Mineral water containing unusual high concentration of boron. Kremna Basin in Western Serbia, Yugoslavia. Radovi Geoinstituta, knjiga 28:107–112 (Serbian, with English summary)
10. EU (1998) EU directive 98/83/EC of 3 November 1998 on the quality of water intended for human consumption. Off J Eur Commun L330: 32–54 05/12/1998
11. EU (2003) EU Directive 2003/40/EC of 16 May 2003 establishing the list, concentration limits and labelling requirements for the constituents of natural mineral waters and the conditions for using ozone-enriched air for the treatment of natural mineral waters and spring waters. Official Journal of the European Union L126/34 22/05/2003
12. EU (2009) EU Directive 2009/54/EC of the European parliament and of the council of 18 June 2009 on the exploitation and marketing of natural mineral waters. Official Journal of the European Union L164/45 26/06/2009
13. Hem JD (1985) Study and interpretation of the chemical characteristics of natural water, vol 2254, 2nd edn. United States Geological Survey Water Supply Paper, Washington, DC, p 129
14. Krajnov SR, Švec VM (1987) Geochimija podzemnych vod chozjajstvenno-pit'evogo naznacenija, vol 235. Nedra, Moskva
15. Landegrin S (1974) Vanadium. In: Wedephol KH (ed) Handbook of geochemistry, vol II-2. Springer, Berlin, 2311
16. Leko M (1901) Hemijsko ispitivanje mineralnih voda u Kraljevini Srbiji. Srpska Kraljevska Akademija, Spomenik, XXXV, Prvi razred, 5:103–157, Beograd
17. Leko M, Ščerbakov A, and Joksimović H (1922) Lekovite vode i klimatska mesta u Kraljevini Srba, Hrvata i Slovenaca sa balneološkom kartom, Ministarstvo narodnog zdravlja, Beograd, pp 141–143
18. Martinović M, Milivojević M (2010) Serbia Country Update, Proceedings World Geothermal Congress, Bali, Indonesia
19. Milivojević M, Perić M (1990) Studija: geotermalna potencijalnost teritorije SR Srbije van teritorija SAP. Rudarsko-geološki fakultet, Beograd
20. Misund A, Frengstad B, Siewers U, Reimann C (1999) Variation of 66 elements in European bottled mineral waters. Sci of Total Environ 243(244):21–41
21. Official Gazette of SFRJ volume 34/79. Regulation on the classification and categorization of groundwater reserves and keeping a data monitoring
22. Official Gazette of FRY volume 42/98. Regulation on the hygienic acceptability of potable water
23. Official Gazette of FRY volume 44/99. Regulation on the hygienic acceptability of potable water
24. Official Gazette of Serbia and Montenegro, volume 53/05 Regulation on quality and other requirements for natural mineral water, Spring Water and Bottled Drinking Water
25. Official Gazette of RS, volume 92/2008. Regulation on way of determination and maintenance of sanitary protection of water supply springs zone
26. Papić P, Krešić N (1990) Specific chemical composition of karst ground water in the Ophiolite belt of the Inner Dinarides in Yugoslavia a case for covered karsts. J Environ Geol Water Sci XV(2):131–135

27. Papić P (1993) Migracija fluora u mineralnim vodama Srbije. Rudarsko geološki fakultet, Univerzitet u Beogradu, Magistarska teza

28. Petrović T, Zlokolica-Mandić M, Veljković N, Vidojević D (2010) Hydrogeological conditions for the forming and quality of mineral waters in Serbia. In: Birke M, Demetriades A, De Vivo B (Guest eds) Mineral waters of Europe. Special Issue, J Geochem Exploration, 107(3):373–381

29. Pokrajac S, Arsenijević M (1977) Hidrohemijska studija termomineralnih voda Socijalističke Republike Srbije. Institut za geološka i rudarska, nuklearna i druga mineralna istraživanja, Beograd

30. Protić D (1988) Zakonomernost raspodele mikroelemenata u mineralnim vodama SR Srbije. Magistarska teza, Rudarsko geološki fakultet, Univerzitet u Beogradu Protić D (1993) Geochemical models of thermal water resources in paleovolcanic calderas of Serbian territory, PhD thesis, Faculty of Mining and Geology, University of Belgrade, (Serbian, with English abstract)

31. Protić DM (1995) Mineralne i termalne vode Srbije. Geoinstitut, Posebna izdanja, knjiga 17, Fond stručne dokumentacije Geoinstituta, Beograd

32. Reimann C, and Birke M (Eds.) (2010) Geochemistry of European bottled water. Borntraeger Science Publishers, Stuttgart, 268 pp. Available online at: http://www.schweizerbart.de/publications/detail/artno/001201002. Last accessed on 18 April 2011

33. Samek K (2004) Unknown quantity: the bottled water industry and Florida's springs. J of Land Use 19(2):569–595

34. Salminen R, (Chief-editor) Batista MJ, Bidovec M, Demetriades A, De Vivo B, De Vos W, Duris M, Gilucis A, Gregorauskiene V, Halamic J, Heitzmann P, Lima A, Jordan G, Klaver G, Klein P, Lis J, Locutura J, Marsina K, Mazreku A, O'Connor PJ, Olsson SÅ, Ottesen RT, Petersell V, Plant JA, Reeder S, Salpeteur I, Sandström H, Siewers U, Steenfelt A, Tarvainen T (2005) geochemical atlas of Europe, Part 1: Background information, methodology and maps. Geological Survey of Finland, Espoo, 526 pp. Available online at: http://www.gtk.fi/publ/foregsatlas/. Last accessed on 18 April 2011

35. Srećković-Batočanin D, Vasković N, Matović V, Erić S (2010) Relics of the ocean crust at the Fruška Gora Mountain Gabbros and Basalts in the Jazak Locality. Proceedings 15th congress of the geologist of Serbia, Belgrade, pp 25–36

36. Teofilović M (1998) Banjske i mineralne stone vode – njihov značaj za život i ekologiju čoveka. Ecologica, vol 5, br 4

37. Teofilović M (1998) Mineralne, termalne i termomineralne voda. Ecologica, vol 5, br 1

38. Vujanović V, Teofilović M, Arsenijević M (1971) Regional study of basic geological, geochemical and genetical properties of mineral waters and spas in Serbia and Autonomous Province Vojvodina. Radovi Geoinstituta, Beograd, Serbian, with English summary

39. Vujanović V, Teofilović M (1980) Problems of geochemistry and genesis of mineral water of Serbia. Radovi Geoinstituta 14(20):123–141 (Serbian, with English abstract)

40. Vujanović V, Teofilović M (1983) Banjske i mineralne vode Srbije. Privredna knjiga, Gornji Milanovac

41. WHO (2006) Guidelines for drinking-water quality. First Addendum to 3rd edn, vol 1 Recommendation. Geneva, p 595

42. Zlokolica-Mandić M (1998) Structural-Tectonic Elements as a Factor in Cave Development. In: Djurovic P (ed) Speleological Atlas of Serbia. Serbian Academy of Sciences and Arts, Belgrade

43. Zlokolica-Mandić M, Papić P (2010) Hidrohemijski atlas podzemnih voda -I faza.Geološki institute Srbije i Rudarsko-geološki fakultet, Univerzitet u Beogradu

44. http://www.mineralwater.rs

Chapter 20
Current Issues and Research Needs for Contaminated Land and Drinking Water Resources Management in Poland

Janusz Krupanek

Abstract The paper presents latest developments for the last 20 years, starting from the 1990s of the twentieth century in management of soil and water resources in Poland with analysis of the needs for further improvements in technological innovations and improvements in management practice. Needs for further improvements concern better administrative procedures, systemic identification of soil and water problems e.g. sites, risk based management and orientation on sustainability in environmental protection, remediation and revitalisation of contaminated sites. Improvements in administrative practices should be based on the experiences gathered so far and good examples from other countries. There is a need for better protection and remediation technologies in terms of their cost-effectiveness and sustainability including application of in-situ techniques in soil and groundwater remediation. Environmental management can also be improved by the development and application of soft tools for problem identification, management and control. Technologies tackling diffuse contamination are also important. Future research topics should concern protecting and enhancing jointly soil functions and water protection especially on built-in areas in accordance with Framework Water Directive and the proposed Soil Framework Directive (COM(2006) 232) framework.

20.1 Water Resources

Water resources in Poland are scarce in comparison with other European countries. The situation might worsen in the future due to climate changes. Water resources in Poland comprise both groundwater and surface water. Available surface water

J. Krupanek (✉)
Instytut Ekologi Terenów Uprzemysłowionych, Katowice, Poland
e-mail: krupanek@ietu.katowice.pl

F.F. Quercia and D. Vidojevic (eds.), *Clean Soil and Safe Water*, NATO Science for Peace and Security Series C: Environmental Security, DOI 10.1007/978-94-007-2240-8_20, © Springer Science+Business Media B.V. 2012

resources amount to 55,143.3 mln m^3 per year (1,400 m^3 per capita). Groundwater resources are estimated at 17.066 mln m3 per year (426 m^3 per capita) [6]. In 2009 10,828.4 mln m^3 of surface water were used and 1,613.8 mln m^3 of groundwater.

For producing potable water and water for industrial purposes the riverine water and groundwater are utilised. The available groundwater resources comprise Quaternary sources with yearly availability 11,293.4 mln m^3 and deeper layers e.g. Tertiary 1,769.3 mln m^3, Cretaceous 2,328.5 mln m^3, and older 1,674.8 mln m^3. The shallow Quaternary layers are in the most part contaminated, threatened by negative impacts or have inadequate for drinking purposes quality.

In 2009 for industrial production processes 7,601.8 mln m^3 of water were abstracted from which 7,331.3 mln m^3 was surface water and 195.6 mln m^3 of groundwater. The surface water is predominantly used for cooling purposes in energy sector.

In principle drinking water is provided to the citizens in Poland according to the requirements of European Union Drinking Water Directive. Yearly 2,067.3 mln m^3 are fed into the line supply system, of which 649, 1 mln m^3 is surface water and 1,418, 2 mln m^3 is groundwater. Most of the Polish population in the cities around 90% is supplied with water through water supply system. The rest of the population, living predominantly in rural areas relies on individual groundwater wells. Use of individual abstraction wells is gradually declining as investments in water supply systems are undertaken by local communities.

Groundwater is abstracted, treated and fed into the supply systems. In similar way surface water is abstracted usually through infiltration wells or directly from water reservoirs, then pre-treated and distributed. Water reservoirs play an important role especially for Polish agglomerations e.g. Upper Silesia. Water reservoirs are threatened by high loads of nutrients and consecutive biological deterioration. Complex protection programs are currently undertaken to improve the situation.

20.2   Impacts and Threads

In general the negative anthropogenic impact is declining as less pollution is coming from the industries and from waste and wastewater management. Basically this is due to investments in municipal infrastructure. The yearly expenditures in water management fixed assets were at the level of 1.7 billion Euros. Positive trend is observed especially in surface water quality. Nevertheless, to secure drinking water supply in a long run it is important to augment water resources in terms of their quality and quantity.

Despite the past efforts, various threats to water resources still remain. They include direct discharges of untreated sewage, industrial wastewater, contaminated land and sediments, urban runoff. As a result surface and groundwater water quality in Poland is still unsatisfactory. According to the state environmental monitoring only 44% of surface water is qualified as very good or good according to European water quality classification (Fig. 20.1).

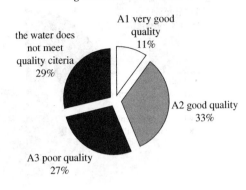

Fig. 20.1   Ground and surface water resources quality in Poland [6]

Biological contamination and inorganic pollution of surface water and ground-water still pose problem in rural areas where water supply through distribution network is not developed and the quality of shallow groundwater used for water abstraction is poor. The problem is aggravated by bad practices in sewage and agricultural wastes disposal. Impact of point and diffuse nutrients emission sources is also important.

Risk of hazardous substances occurrence in the environment has to be taken into account although it is important mainly in industrialized and urbanized areas. The problems are of local character and they are limited to areas impacted by relevant industrial activities or contamination sources. They are related to soil, groundwater or sediment contamination. In these cases the raise of water supply costs is observed along with high energy demand and other negative social and environmental aspects.

As an example decline in mercury and cadmium concentration levels in environmental media during the last 20 years in Poland confirms the improvement in the water quality but in recent years the concentrations stabilize at acceptable level (Fig. 20.2).

Point emission sources of these substances are less important today because of the EU directives and implementation of appropriate activities in the industrial and municipal sectors.

There is visible change in the impact on water resources, with improvement in point sources environmental management and the pronounced role of diffuse sources. The graph presents (Fig. 20.1) prognosis of mercury load in Upper Silesia Kłodnica river catchment located in highly industrialized region (the EU project SOCOPSE, information available at www.socopse.se).

Diffuse pollution sources are recognized as those having today the biggest impact on drinking water resources. They include contaminated land and groundwater,

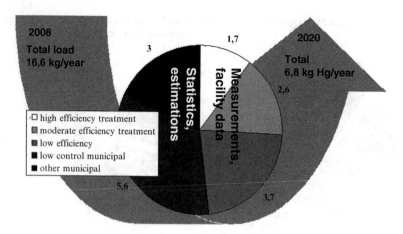

Fig. 20.2 Long term prognosis of mercury contamination decline in Kłodnica river basin (results of SOCOPSE project)

contaminated sediments, soil, urban run off, common consumption, residual wastes, landfills, household activities (e.g. dioxins as result of backyard burning, washing of textiles). Soil contamination is an essential source of hazardous substances. It is connected with post-military sites, airfields, terminals, distribution network and fuel depots Industrial facilities, landfills, port areas.

It should be underlined that from the country perspective the water resources are sufficient today. Aggravated problems with quality of drinking water resources can be important only on local scale. In the past, degradation of water resources lead to utilization of other resources (e.g. transport of water from resources located in clean area, switching between the resources) or through application of water pre-treatment (in case of surface water).

20.3   Water Resources Management

Today, water policy framework in Poland is based on European Union Water Framework Directive (WFD) which requirements are paramount in river basin management. According to WFD, the water related management is based on sound and publicly accepted environmental goals, integration of surface and groundwater issues, holistic perspective, economical approach with real costs of water services, environmental liability based on polluter pays principle. Decisions should take into account rationality of goals, cost-effectiveness of measures, integrate various aspects of the management and should be a result of stakeholder's involvement.

River basin management plans in Poland were officially established by Polish government in 2011. They set priorities in water management with the focus on municipal, industrial and agricultural sectors. The plans are in line with the developments being undertaken in Poland in the past 20 years during Poland's transformation period when high investments in municipal infrastructure and implementation of environmental systems were undertaken. These actions performed under environmental policy in Poland have improved essentially the quality of rivers. The national efforts were supported by the available EU structural funds.

Along with the river basin planning procedure, basic settings of the management are established. Environmental quality assessment is based on EU regulations setting the environmental quality criteria. Water Quality Standards EQS for surface and groundwater and selected inorganic and organic substances and drinking water standards are established along with discharge limits for selected water related activities. The EQS for surface water are set for hazardous substances, nutrients and major inorganic constituents including metals. Groundwater quality indicators are set for monitoring and reporting purposes. There are also set soil quality standards taking into regard three land functions.

The impacts on water quality are controlled by water related regulations and policies concerning environmental management in industry, waste management and soil management. The industrial impact on water resources is regulated by Integrated Pollution Prevention and Control (IPPC) Directive requirements concerning industrial branches having the biggest environmental impact. Strict requirements are imposed on municipal wastewater treatment plants for effluents quality and the quality of inflow from selected industrial activities. To some extent the small and medium enterprises as less controlled and can potentially have important environmental impact.

Soil management is based on polluter pays principle as well. The site owner is obliged by Environmental law to remediate the sites. Recently imposed IPPC regulations also oblige industries to provide reports on soil quality.

The national government and its agencies are responsible for the abandoned sites having the highest environmental risks. Successful state programs were and are carried out, starting from the 90-ties and they focus on post military sites, airfields, munitions sites, pesticide dumping sites, and selected hot spots e.g. chemical factories. Another important issue is waste processing and management of old landfill sites.

Taking into account the whole regulatory framework and the implementation by the addressed actors the policy was successful with eliminating the main risks to water resources especially those connected to point sources, both industrial and municipal. The legal system is in principle in place but development of practice can bring further improvement.

20.4 New Challenges

Sustainability of the water supply is the new challenge to improve the cost-efficiency and to provide variety of benefits from societal point of view [4].

Taking into account the results of research, experience and forecasts of the societal development and economical growth new challenges in the water sector arise:

- High impact of not controlled diffuse pollution including agriculture,
- New contaminating substances e.g. endocrine disruptors including pharmaceuticals,
- Climate change threats e.g. aridification in some parts of Poland,
- Expected rise in costs of public water supply,
- Spatial changes and negative geochemical processes,
- Local and regional water resources deficiencies,
- Integration and augmenting of water related soil functions
- Augmenting of water resources,
- Rehabilitation of degraded waters

Long term securing of water resources is important from local development perspective in regions where problems are especially aggravated e.g. industrialized mining areas with highly changed hydrological and hydro geological conditions.

From public perspective the availability and quality of water is very important. The cost of water services is recognised as the key factor by stakeholders including citizens, as the full price for water delivery and sewage discharge has to be covered by the water users. Cost – efficiency and energy – efficiency pose a new challenge on the technological development as well. Furthermore, full recognition of the threats from various pollution sources to water resources, including groundwater is needed. These challenges require more action in integrating water and soil management.

20.5   Innovation and Environmental Technologies

Further improvements in management of water resources require better use of existing technologies, development of new methods and approaches [4]. The purpose can be served by linking science, technology, policy and practice. For that reason European Union Environmental Technology Action Plan (ETAP) was adopted by the Commission in 2004 [2] to cover a wide range of activities promoting eco-innovation and practical use of environmental technologies. One of the main goals of the plan is to strengthen the opportunity for environmental technologies development and their market implementation. It was followed by establishing Polish Environmental Technology Action Plan [14].

Reflecting the European policy [2, 3, 5] a voluntary initiative was established in 2005 [17] as Polish ETAP Polish Platform of Environmental Technologies (PPET). The main role of the initiative is to match stakeholders involved in research, development and implementation of environmental technologies with an aim to boost eco-innovation in Poland. Strategic Research Program was developed in down – top approach in relation to practical needs identified in Poland and EU policy demands. The knowledge gaps were identified by Polish research community. Priority technological areas in water and soil management were selected.

Basically, in water and soil management practice the commonly used technologies and approaches prevail. As the new challenges arise needs for further improvements are foreseen both in supporting the administrative procedures with soft management tools of assessment, monitoring and control and in application of remediation and treatment methods.

In soil remediation the universal standard management procedures and remediation techniques are applied. The selectivity of the management techniques towards the site unique characteristics is low. Use of soft tools (organisational) in the decision processes of soil, surface water and groundwater protection management is inadequate [17]. It concerns better understanding of soil and water problems, selection of the technologies and strategies. In many cases this approach leads to insufficient remediation or excessive costs. It is important especially in management of difficult and complex cases.

Basic technologies for cleanup of contaminated soil and groundwater (pump and treat, dig and dump) prevail as they are recognized by the practitioners as the most reliable and efficient in practical application. More attention can be put to more selective application of already existing new technologies, smart coupling of technologies and research on emerging technologies [17]. It is especially important in the view of green remediation and sustainable remediation concepts based on holistic view of the environmental, social, and economical perspectives [7]. The existing barriers to application of innovative technologies relate to reliability, uncertainty concerning the costs, administrative procedures [21]. An unfavourable condition exists for uptake of new technologies and barriers such as public procurement.

It should be noted that for the time being the baseline technology development and application is sufficient, having in regard available resources and administrative practices. Nevertheless, the new challenges require new technologies and new thinking.

20.6 Water Related Technologies

In water sector innovative cost-efficient technologies for contaminants removal in treatment processes needs to be further developed [17]. The membrane technologies (e.g. nano filtration and reverse osmosis) in water and wastewater treatment are highly relevant in the context of hazardous substances and emerging substances removal from wastewater streams [20]. Currently, this technology is not used in municipal wastewater treatment, only in selected industrial applications. Use of this technique is potentially desirable in wastewater treatment plants and in certain industrial processes. It is the more interesting as the membrane technologies can be used for upgrading existing wastewater treatment plants with many additional benefits such as reduction of space (e.g. replacement of retention tanks). Further improvements should be focused on specific demands, wastewater characteristics and cost-efficiency.

In industry the remaining challenge is improvement in closed water cycles in production processes both saving the water and reducing environmental impact. The technologies are branch specific and tailored solutions are required. Good examples exist and they show that this solution can be potentially a good option not only as technique for contaminants removal but for saving the water use through recycling the treated water.

One of the basic technical problems in Poland is the removal of nutrients at municipal wastewater treatment plants. Techniques for reducing biogenic elements in discharge to surface waters have to be addressed. In 2009 around 1.2 thousand tons of phosphorous and 21.1 thousand tons of nitrogen was discharged from municipal treatment plants [6]. Cost – efficient removal of phosphorous and nitrogen is required as more stringent regulations are put into practice. Denitrification, chemical precipitation and prolonged time retention are the possible options. These are especially important for refitting existing facilities.

It is important to tackle the diffuse sources of nutrients and hazardous substances. Agriculture in Poland is an important sector which will be intensified in future, with development of large farming facilities and intensive use of pesticides. Beside better methods of husbandry and cultivation, a protective measure has to be applied with constructed wetlands, buffer zones, retention ponds. Bioengineering systems in water protection are a good solution as it can be for example combined with biomass production. Buffer zones and wetlands preventing biogenic elements flow into surface water system can be also efficient solution in augmenting water resources.

Efficient water management system requires good monitoring and control tools. It is the more demanding that specific substances have to be monitored. Monitoring of the environment (monitoring, prevention, micro-contaminants, soil and water environment) will need development new sampling techniques (e.g. passive sampling) and analytical skills.

Water management should be integrated with many activities in industrial sector, municipal sector, spatial planning, and housing and development strategies. Integration needs knowledge and competences and modelling tools. Models of integrated water management in catchment scale are important to serve the purpose. These models concern surface runoff, retention, groundwater alimentation and water abstraction and are needed in spatial planning [12]. Another important set of tools is management of the infrastructures including mass transport in sewerage systems, storm overflows for optimizing the water infrastructure and management of the processes.

Losses of water during transport require water quality conservation. As the infrastructure is aging high costs are to be expected. Models of operation and revitalization of water infrastructure (on-excavation methods, water supply systems, sewage systems, wastewater treatment plants, hydro technical constructions) should be widely applied. The new methods gradually are introduced in practice.

20.7   Soil Related Technologies

In general, site remediation requires strategic approach incorporating green/sustainable contaminated land remediation rules [15, 18]. This can be translated into improvements in administrative procedures and remediation practice [17].

Heavy metals contamination is a risk factor for long term soil quality control in some regions in Poland. This is especially important in the mining areas impacted by historical activities of zinc ore smelting. The areas are of large spatial extent and with heavy metals concentrations above the allowable level. Site management in these cases requires in the first place reduction of environmental or human health hazards connected with agricultural or recreational land functions. For these areas cost-effective long term risk reduction measures are needed. The promising techniques comprise: phytostabilisation of heavy metals in soil and application of evapotranspiration cover for landfills as an isolating barrier for contaminants infiltration down groundwater. It tackles also the issue of developing knowledge and practical applications of artificial soils.

In Poland sites contaminated with hydrocarbons petrol, jet fuel diesel oil, and chloroorganics: trichloroethane, trichlorethane, tetrachloroethane pose essential risk to ground and surface water resources. There are numerous examples of threats to groundwater resources which were successfully resolved or which still have to be tackled [11]. Among the contamination sources there are post-military sites, airfields, terminals, distribution network and fuel depots industrial facilities, landfills, port areas, dry cleaners. Innovative in-situ technologies for bioremediation of soils contaminated with organic compounds is an interesting option to limit the risks in rehabilitation of the environment and land revitalization [19]. For cleaning contaminated soil ex-situ and off site bioremediation techniques are frequently used in Poland. The ex-situ bioremediation was under development and testing within scientific projects e.g. project done in cooperation with US EPA in Czechowice – Dziedzice petroleum plant: ground and groundwater contamination with acidic tar with testing of bio pile bioremediation techniques [16]. Combination of in situ methods, including hydrofracturing, bioventing and steam injection for dealing with jet fuel contamination was initiated at an air base at Kluczewo located in watershed of Miedwie Lake which has major recreational value and is used for water supply to half a million people [1].

Technologies and approaches that optimize in-situ soil and groundwater remediation should be tested for cost, time and sustainability. Sites contaminated with petroleum hydrocarbons and chlorinated hydrocarbons are being considered as the sites most relevant for in-situ technologies. One of the current research is done in the UPSOIL project (information available at www.upsoil.eu) financed by EU 7th FP. The project aim is focused on coupling various technologies: injection technology with knowledge on optimization chemical and biological treatment to provide efficient remediation approach.

For pump and treat technologies improvements in contaminants removal techniques are important mainly to lower the cost with regard to high efficiency of the process. Organic sorbents for removal of organic contaminants, including breakdowns is one of the promising approaches which are identified as priority of innovation development.

Site characterisation and monitoring are important elements in preparing remediation, selection of cleanup methods and taking up the remedial decisions. Tools for environmental characterisation and monitoring at contaminated sites such as GIS, teledetection, non-invasive technologies, sensors, etc. can bring new dimension to remediation process. In the decision making process modelling tools

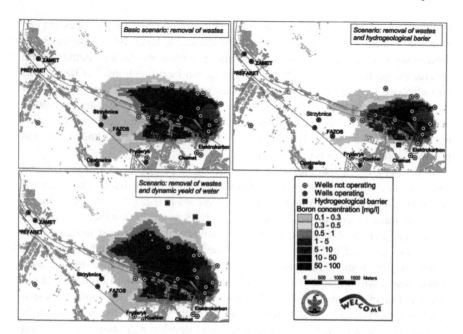

Fig. 20.3 Use of modelling in developing scenarios for boron plume developments in Tarnowskie Góry case of chemical plant contamination [13]

integrated with site characterization a monitoring can be a useful tool [9, 13]. Good site characterization is crucial especially for monitored and enhanced natural attenuation. For example analyses of various plume development scenarios made on the basis of contaminant flows in groundwater (WELCOME project information available at www.euwelcome.nl) at Tarnowskie Góry mega-site contaminated through chemical production with inorganic (boron, barium) and organic components, (tetrachloroethen) proved the value of modelling tools (Fig. 20.3).

Administrative decisions concerning land remediation are based on strict requirements. Insufficient use of risk assessment and management (environmental, health) slows down the remediation as it leads to excessive costs [10]. Currently modelling tools are used in complicated cases for sites with hydrocarbons LNAP and inorganic.

In Poland it is observed that new characterization techniques researched by scientific institutions – are gradually coming into practice. These include: direct push technique, on-line monitoring, modelling, Isotopes techniques, hydrolic tests, georadar. The latest is implemented in day to day practice.

20.8   International Cooperation

The challenges in securing water resources are universal for European countries although different issues are pronounced in particular countries. The problems in many cases can and should be solved through international cooperation. The potential areas comprise:

- Wide communication and knowledge transfer,
- Exchange of administrative and technical experiences,
- Joint research and technology demonstration

In Europe and on international level knowledge and experience concerning protection of water resources is accumulated. It concerns water treatment processes, augmenting of water resources, soil protection and remediation of contaminated sites. Exchange of administrative practice might be very useful as the EU countries have various approaches and experiences. In Poland it can help to improve administrative procedures. This concerns use of proportionality principle, degraded site identification and assessment, decision making and control procedures, use of natural attenuation concept. It can help with understanding of barriers in administrative practices and legal gaps, and overcoming the lack of experience in application of innovative solutions. The accumulated knowledge on remediation and the results of EU projects can help in that matter. In 2010 project: Tailored Improvement of Brownfield Regeneration in Europe – TIMBRE financed by 7th EU Framework Program was launched to strengthen the strategic thinking in site remediation (information available at: www.timbre-project.eu). Its aim is to develop tools for holistic and site specific approaches to contaminated sites remediation.

In that respect, good example is the cooperation between German and Polish researchers and practitioners within TASK initiative funded by German Ministry of Science and Education.

The cooperation concern knowledge transfer based on German scientific projects and their practical application. One of the topics is knowledge transfer on monitored natural attenuation approaches to its practical application in Poland. Workshops and dissemination of materials are provided for representatives from Regional administration, scientific institutions, consultants, technology providers. Another example is Central Europe Program project COBRAMAN (www.cobraman-ce.eu) which intention was development of competences of local authorities in contaminated site management.

Another important issue for sharing the experiences is nutrient and hazardous substances and reduction of their emission to surface waters in river basin management. Strategic approach in this area needs improvement. Strategic assessment of hazardous substances inputs to river basins was undertaken in EU funded projects with recommendations on cost-efficient technologies and approaches in emission reduction of hazardous substances. In 7th European Union Framework Program SOCOPSE project focused on development of strategic management system for preparing river basin management plans in the aspect of priority substances.

Another project VISCAT (http://www.cs.iia.cnr.it/EUROCAT/ project.htm) aimed at evaluation of nutrients and selected contaminants inputs and defining reduction strategy in Vistula river basin. Project "Control of hazardous substances in the Baltic Sea region" (COHIBA) of Baltic Sea Region Program focused on development of strategy for Baltic Sea region of reducing the HELCOM hazardous substances (http://www.cohiba-project.net).

Strengthening the link between science and business on international level is carried out through technological platforms – The European Water Platform (http://www.wsstp.eu) in the water sector and NICOLE (www.nicole.org) and EUGRIS (www.eugris.info) in soil remediation. Technology development initiatives of research communities: SOILTech cluster (information accessible at http://www.soiltechnologyresearch.eu/) provide platforms for information exchange and scientific networks – ENVITECHNet (www.envitech-net.org).

To develop trust in innovative technologies verification schemes can bring a new stimulus. For that European initiative on Environmental Technology Verification (ETV) scheme is encompassed by international community (information accessible at http://www.eu-etv-strategy.eu). Polish organizations took part in a set of projects focusing on development of criteria assessment PROMOTE project: http://www.promote-etv.org/.

20.9   Conclusions

Further socio-economical development needs more specific and integrated management of water resources. This has to recognize various aspects of management in industry, municipal sector development, spatial planning.

Technological development should not only be focused on hard technologies but predominantly on know how transfer and competences development of the various stakeholders on efficient application of the innovative approaches in practice.

Sustainable and green approach in protection of water resources is the major future challenge. One of the key issues is cost-efficiency and broader consideration of social and environmental benefits and impacts. Rising energy costs, spatial constraints and more stringent environmental requirements will need changes in comparison with the current practices and the technological state of the art.

In Poland in many cases the innovative technologies are available but they are too costly in practical application or they are characterized with negative characteristics like production of hazardous wastes, high energy or material demand. These aspects of the technologies have to be improved with regard to specific contexts of their application.

Preserving and augmenting the water resources is one of the key topics. For this purpose better characterization and management of water resources has to be implemented in practice. In the context of augmenting water resources strong focus has to be put on soil functions.

It can be expected that societal needs and sustainability constraints will require gradual changes in the infrastructures and water management systems with better management and control and gradual improvements in the infrastructures. In water sector specific areas of technological development arise including wastewater treatment techniques – new more cost- efficient, energy efficient and sustainable approaches.

In soil sector there is a need for wide implementation of good practices in planning of remediation process with tailored solutions and more detailed and reliable characterization and monitoring of contaminated sites.

A hard technology needs to be developed and especially adapted to practical applications. In-situ methods are especially interesting for research and practical implementation. Very promising research area is coupling of technologies within optimized remediation strategies. Emerging remediation methods and those which are relatively new in Poland should be recognized in practice.

It should be underlined that use of the techniques and technologies should be wise and tailored. It requires development of competences potential of implementation involvement of firms and environmental technology verification.

References

1. Benoit Y, Haeseler H, Slack W, Broeker T, Klind K, Nilsson B, Kasela T, Tsakiroglou T (2005) Stimulating soil decontamination within a fractured soil at Kluczewo Poland Session D pp 1654–1661 ConSoil 2005 CONSOIL 2005, Proceedings of the 9th international FZK/TNO conference on soil-water systems, 3–7 Oct 2005, Bordeaux.
2. CEC (2005) Commission of the European communities, communication from the Commission, Report on the implementation of the environmental technologies action plan in 2004 Brussels, 27.1.2005 COM (2005) 16 final
3. CEC (2007) Communication from the commission to the council, The european parliament, The european economic and social committee and the committee of the regions, Report of the environmental technologies action plan (2005–2006) Commission of the european communities Brussels, 2.5.2007 COM (2007) 162 final
4. CMRP (2008) Council of ministers republic of Poland (2008) The national environmental policy for 2009–2012 and its 2016 outlook. Ministry of the environment Warsaw 2008 http://www.mos.gov.pl/g2/big/2009_07/2826c539c3015384e50adac8fe920b0b.pdf
5. COM (2004) 38 Final communication from the Commission to the council and the European parliament, stimulating technologies for sustainable development: An Environmental Technologies Action Plan for the European Union, Brussels, 28.1.2004
6. CSO (2010) Central statistical office (Główny Urząd Statystyczny), Environment 2010, Warsaw 2010, Statistical information and elaboration
7. Ellis DE, Hadley PW (2009) Sustainable remediation white paper—integrating sustainable principles, practices, and metrics into remediation projects, 2009 U.S. sustainable remediation forum, Published online in Wiley interscience, pp 5–10 (www.interscience.wiley.com). REMEDIATION Summer 2009. DOI: 10.1002/rem.20210
8. http://ec.europa.eu/environment/enveco/eco_industry/pdf/ecoindustry2006.pdf
9. Krupanek J (2005) Integrated management strategy for risk reduction of groundwater contamination at Tarnowskie góry megasite, NATO/CCMS pilot study: Prevention and remediation issues in selected industrial sectors: Mega-Sites Ottawa, Canada, June 12–15, 2005. http://www.cluin.org/ottawa. Accessed 2011

10. Krupanek J (2006) Barriers and bridges to risk assessment and management of contaminated sites in urban areas: Upper silesia case study, NATO/CCMS pilot study: prevention and remediation issues in selected industrial sectors Athens, Greece June 4–7, 2006. www.cluin.org. Accessed 2011

11. Krupanek J (2007a) Inventory of potential contaminated sites in upper silesia region report of the pilot study meeting prevention and remediation in selected industrial sectors sediments, Ljubljana, Slovenia, June 17–22, 2007, Annual report No. 281 EPA 542-R-07-014 Aug 2007, Poland, pp 6. www.cluin.org. Accessed 2011

12. Krupanek J(2007b) POPS reduction strategy in surface water of industrialized regions—Klodnica river case study report of the pilot study meeting prevention and remediation in selected industrial sectors sediments, Ljubljana, Slovenia, June 17–22, 2007, Annual report No. 281 EPA 542-R-07-014 Aug2007, Poland, pp 24. www.cluin.org. Accessed 2011

13. Krupanek J, Janikowski R, Korcz M(2003)Concept of sustainable development for Tarnowskie góry megasite – Poland. Session D PP Proceedings 3005–3009 in ConSoil 2003 May 12–16, 2003, Gent, Belgium, (the 8th International FZK/TNO Conference on contaminated soil)

14. ME (2006) Roadmap for implementation of environmental technology action plan in Poland ministry of environment, Warsaw 2006

15. NICOLE (2010) NICOLE Road map for sustainable remediation Sept 2010, Network for industrially contaminated land in Europe, www.nicole.org, Accessed 2011

16. Plaza GA (2006) Bioremediation of petroleum contaminated soils from the refinery by biopile Prace Naukowe Instytutu Inżynierii Ochrony Środowiska Politechniki Wrocławskiej. Monografie 2006, vol 80, nr 47, s.141

17. PPTŚ (2005) Program Badań Strategicznych Polskiej Platformy Technologii Środowiskowych IETU. Accessed in 2011 at www.ietu.katowice.pl

18. Raymond RL. Jr, Dona LC, Haddad EH, Kessel LG, McKalips PD, Raymond ChN, Vaské J (2009) Description and current status of sustainable remediation, 2009 U.S. sustainable remediation forum, published online in Wiley interscience (www.interscience.wiley.com). REMEDIATION Summer 2009, pp 11–33. DOI: 10.1002/rem.20210

19. SPIG (2003) Environmental technologies action plan Discussion paper report. Report from the soil protection issue group as a contribution to the environmental technologies action plan July 2003, Accessed 2011 at http://ec.europa.eu/environment/etap/pdfs

20. WIG (2003) Environmental technologies action plan, discussion paper report from the water issue group as a contribution to the environmental technologies action plan, Aug 2003, Accessed 2011 at http://ec.europa.eu/environment/etap/pdfs

21. Ernst and Young (2006) Eco-industry and its size, employment, perspectives and barriers to growth in an enlarged EU. Final report for the European commission DG environment, Sept 2006

Chapter 21
The Role of the Regulator in the Water Management in the Czech Republic – Case Study of a Large Remediation Project

Kvetoslav Vlk, Jaroslav Zima, Zdena Wittlingerova, and Jirina Machackova

Abstract In the Czech Republic the universal objective of the national water management policy is to create conditions for sustainable management of the finite water resources. This implies that all forms of water resource use should be in compliance with water and aquatic ecosystem protection requirements while applying measures to reduce the harmful effects of contaminated water. The key principles of the water management policy are derived from the EU Water Framework Directive, other water management directives and the renewed EU Sustainable Development Strategy. The Water Protection Department of the Ministry of the Environment is the central water management authority. The Czech Republic also pays a big attention to environmental damages mostly on ground water aquifers created before privatization of industry, also partly of agriculture production and after Soviet Army stay (1968–1991). A case study of remediation of a large-scale petroleum contamination of soil and groundwater at the former Soviet Army air base at Hradcany is reported. The air base and its vicinity are part of the sedimentary complex of the Bohemian Cretaceous Basin, which is a significant groundwater supply source for

K. Vlk (✉) • J. Zima
Ministry of Finance of the Czech Republic, Letenska 15, Prague, Czech Republic

Faculty of Environmental Science, Czech University of Life Sciences,
Praha 6, Suchdol, Czech Republic
e-mail: kvetoslav.vlk@mfcr.cz

Z. Wittlingerova
Faculty of Environmental Science, Czech University of Life Sciences,
Praha 6, Suchdol, Czech Republic

J. Machackova
AECOM CZ Ltd, Trojská 92, 171 00, Prague 7, Czech Republic

F.F. Quercia and D. Vidojevic (eds.), *Clean Soil and Safe Water*, NATO Science for Peace
and Security Series C: Environmental Security, DOI 10.1007/978-94-007-2240-8_21,
© Springer Science+Business Media B.V. 2012

the Czech Republic. The site served as an army air base from World War II to 1992 and a large contamination plume – 28 ha 7,000 metric ton of petroleum products release –had to be remediated between 1997 and 2010.

21.1 Site Characteristics

The site was used for military purposes from 1940 to 1991. These military activities resulted in extensive contamination of the soil and groundwater by petroleum products (70% is represented by jet fuel) in an area of 28 ha (Figs. 21.1 and 21.2). The site's subsoil consists in Quaternary sandy-gravel river terrace deposits (thickness 4 m), underlain by Cretaceous marine sediments, consisting of weathered medium-grain sandstone approximately 50 m thick. The site is a part of the Bohemian Cretaceous Basin, the most important source of high quality groundwater in the Czech Republic. The endangered aquifer is the only source of drinking water in the region and the presence of extensive contamination limits future uses and revitalization of the

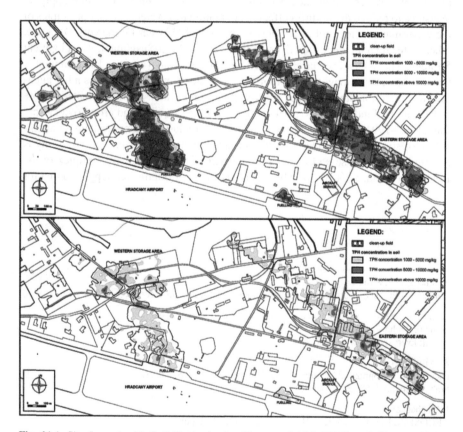

Fig. 21.1 Site Layout with Soil Contamination Extent – Initial (1997) and Closure (2009) Conditions

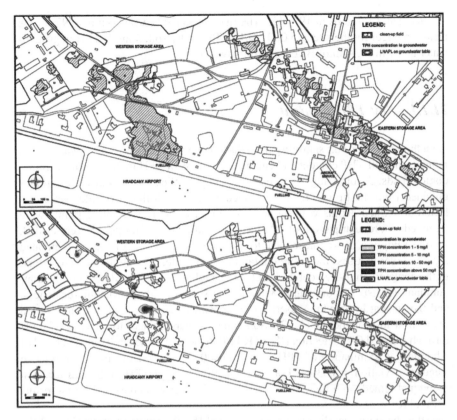

Fig. 21.2 LNAPL and Groundwater Contamination Extent – Initial (1997) and Closure (2009) Conditions

site. Three principal contamination source zones were defined – the so-called eastern and western delivery and storage areas and a fuelling area (Figs. 21.1 and 21.2).

The contamination migrated from the source zones in the direction of the groundwater flow to the river and formed two main plumes: the eastern plume and the western plume. The eastern plume is present in unconsolidated sandy Quaternary river sediments, which represent probably a former riverbed. The western plume, which was created by connection of the plumes originating at the fuelling area and the storage area, is present in the consolidated Cretaceous sandstone bedrock [9, 10]. The thickness of the contaminated soil layer varied between 1 and 8 m; in the source areas both unsaturated and saturated zones were contaminated, but in secondary areas, which were contaminated by migration of free oil phase, only a layer around the groundwater table (GWT) was contaminated - the so-called smear zone extending up to 2 m below GWT. Jet fuel acted as a typical light non-aqueous phase liquid (LNAPL) at the site. The maximum contamination depth was driven by the GWT level and varied between 5 and 10 m, with maximum contamination levels (11,000 mg/kg) present within the range of groundwater fluctuation [9, 10]. The total amount of TPH, which was released into soil and groundwater at the site, was estimated at 7,150 tons in 1997 [5]. The risk assessment carried out in 1997 set the clean-up target for the site at 5,000 mg/kg TPH in dry soil, 5 mg/L TPH in groundwater and 1 mg/L sum of BTEX in groundwater.

21.1.1 Soil TPH Content

A method based on soil sampling from discrete depth intervals (0.5 m) was used to monitor the soil THP content and calculate the mass of TPH in the soil of the remediation fields. A spiral auger was used for sample collection with discrete samples obtained from the tip of the auger immediately after drilling. Samples were collected at defined depths with an accuracy of ±15 cm. The sampling points were arranged in a regular pattern of 20 m × 20 m and 20 boreholes per ha were drilled. During sampling, it was necessary to evaluate the thickness of the contaminated layer in order to assess the entire contamination profile, i.e. sampling had to extend to the contamination-free depth. Point soil sampling was selected because the contaminant consists of dominantly volatile and semi-volatile petroleum substances, and the standard procedures used for compositing of samples (quartering) would result in a loss of the VOCs [2]. Soil samples were subject to analysis of TPH through infrared (IR) detection (ISO TR 11,046 method). The confidence interval of laboratory determination of TPH-IR, with $\alpha = 0.05$ expressed as twice the relative standard deviation, was 30%. For mass quantification of TPH in the soil for each individual remediation field, the TPH-IR soil data set was converted into a results matrix [10].

After the statistical data had been analyzed [5], the trimmed mean showed the best correlation in all the samples of the entire Hradcany site. The calculation of mean values was based on the trimmed average for $\alpha = 0.15$, which disregards 7.5% of the smallest and 7.5% of largest sample values [3]. The sample data have been used for the calculation of the trimmed average and the confidence interval ($\alpha = 0.1$) of the TPH content in unsaturated and saturated zones of an individual clean-up field [10]. The obtained average was then multiplied by the remediation field area, the sampled thickness of the contaminated layer and the soil bulk density (1,650 kg/m^3). The result is an estimate of the mass of petroleum contamination (in tons) in the remediation field. The field specific value was calculated by dividing the value by the field area; the accuracy of value is in range ±25–30%.

The complex remediation of the site based on a combination of several remediation technologies started in 1997, when *in situ* technologies were gradually applied on the site. Removal of volatile organic compounds (VOCs) via soil vapor extraction (SVE) and air sparging (AS), light non-aqueous phase liquid (LNAPL) vacuum extraction, and aerobic TPH biodegradation supported by oxygenation via air sparging and application of nutrient solutions have been applied on the site [11]. The clean-up had two phases – first phase was focused on maximum removal of LNAPL (vacuum extraction), second phase was aimed on optimization of favorable conditions for aerobic biodegradation in the entire contaminated profile (air sparging + nutrients). The contaminated area was divided into individual remediation fields

Fig. 21.3 A bailer with oil and water phases

with areas of 0.5 – 4.0 hectares for the purpose of remedial operations and the clean-up technology was consequently installed in each of these contaminated areas.

21.1.2   Groundwater TPH Content

Groundwater was regularly sampled in a year interval in venting wells. Measurement of the presence of free oil phase on the groundwater table was the first step of groundwater sampling. The presence was detected by a bailer (see Fig. 21.3) and if the phase was present, the thickness was measured and further groundwater sampling was not performed. If no detectable phase was present a groundwater sample was taken. Prior to the collection of groundwater samples, groundwater table level (GWT), pH and temperature were measured. Groundwater samples were collected with a peristaltic pump in dynamic state, after stabilization of the monitored parameters (pH, temperature). Groundwater samples were subject to analysis of TPH through infrared (IR) detection (ISO TR 11,046 method). The oil phase presence was monitored with density 100 wells per hectare, GW TPH content was monitored in 25 wells per hectare.

21.1.3   Map Construction

For the statistical analysis, the spatial distribution of data was neglected and the data were transformed into a one-dimension data file. Further interpretation of the TPH content in the soil was based on two-dimensional maps showing the spatial layout of contaminants. These maps were based on sampling points with geodetic coordinates obtained with an accuracy of 1 cm (in S-JTSK and Balt systems). Data on TPH concentrations were used to calculate a one-side trimmed mean of values greater than 1,000 mg/kg of TPH from layers of the saturated zone (zone II, see Table 1). [7] The basic graphic interpretation was produced with Surfer 6 software with Kriging selected as the interpolation method. The computer output was optimized by hand to produce a final map, which enables to evaluate spatial changes of TPH distribution in clean-up fields during the clean-up.

21.1.4   Measurement of the Mass of Petroleum Hydrocarbons Removed by Remediation Technologies

Remediation monitoring is specific to the technology applied. The mass of removed petroleum contamination is assessed by evaluating VOCs removed through SVE, the mass balance of biodegradation calculated using respiratory gases (O_2, CO_2) monitoring data and free product mass removed via vacuum extraction.

The mass of VOCs removed through SVE/AS is quantified using the concentrations of VOCs in the extracted air volume [1], and the mass of biodegraded petroleum hydrocarbons is calculated using concentrations of respiratory gases in the extracted soil air. The metabolized hydrocarbons are stoichiometrically quantified from the volume of oxygen consumed in the remediation zone using Eq. 21.1 [1, 8]. The production of CO_2 is monitored as well, for control purposes [8]. As the mass estimation based on carbon dioxide production is influenced by soil carbonate cycle (precipitation as carbonates) and also by incorporation of carbon to microbial biomass, observed values are usually slightly lower, but proportional to those derived from oxygen consumption, with exception for highly alkaline soils, where gaseous CO_2 is buffered [8]. At the test site, the ratio of the mass estimation based on carbon dioxide production to the estimation based on oxygen consumption was 91–100%, with average 93%, which indicates efficient proportion between O_2 consumption and CO_2 production and negligible oxygen consumption for chemical oxidation of reduced species of iron and manganese [4]. The respiratory balance was further adjusted using background respiration levels, which were quantified on the basis of the remediation field R operation in a non-contaminated area for a period of 2 years. The background respiration was 2.3 ton of organic substances/ha/year and this amount was subtracted from the biodegradation mass balance calculation in the clean-up field area [5].

Measurements used for the mass of removed VOCs and biodegradation were conducted at monthly intervals and the calculation of the total mass of removed contamination was based on results extrapolated between the measurement dates.

$$B_c = ((Q \times (20.9 - C_{RP}) / 100) \times \rho_{RP} \times C) \qquad (21.1)$$

B_c Total balance of biodegraded hydrocarbons in kg/day for a given remediation field

Q Ventilator flow rate in m³/day

C_{RP} Respiratory gas concentration (O_2, CO_2) in volume %

ρ_{RP} Respiratory gas specific weight (O_2 1.429 kg/m³, CO_2 1,950 kg/m³, regular conditions)

C mass ratio $C_{10}H_{18}/O_2$, 0.29, $C_{10}H_{18}/CO_2$, 0.31,

based on stoichiometric equation

$C_{10}H_{18} + 14.5O_2 \rightarrow 10CO_2 + 9H_2O$, hydrocarbon formula derived from complete qualitative analysis of LNAPL at the site

The accuracy and representativeness of the mass balance was estimated by the measurement error. Calculations were based on the measured volume of extracted air and the concentrations of VOCs and respiratory gases. The extracted air volume was obtained by multiplying the streamline flow velocity in the SVE system, measured with a hot-wire anemometer (TA4/5, Airflow), by a known cross-sectional area of the SVE piping. The air volume measurement error was determined from the anemometer relative error, which was 5%. VOC contents were monitored through TPH-GC laboratory analyses utilizing air sorption on an active carbon filter (NIOSH −1,501 methodology). The confidence interval for the VOC analysis with $\alpha = 0.05$ was 15%. Concentrations of respiratory gases were measured in the venting piping with a portable multi-gas analyzer (Anagas 95, GI). The relative error of respiratory gas concentration measurements was 2.5%. When calculating the VOC and biodegradation mass balance, the measured values were multiplied and the mass balance error was obtained as a summation of the relative measurement errors. The confidence interval ($\alpha = 0.05$) was determined as double the measurement error, which was 25% in the case of the mass of removed VOCs and 15% in the case of the biodegradation mass balance.

The mass of TPH removed through free phase vacuum extraction was derived from direct measurement of the volume of oil removed using calibrated containers, as well as TPH contamination levels in water after gravimetric separation. The estimated error of this mass estimate was 20%.

21.2   Results

21.2.1   Clean-Up Results in a Selected Clean-Up Field (I, 1.4 ha)

Remediation of Field I started in 1998. SVE/AS and vacuum extraction of LNAPL from the GWL were used for the first 18 months of remediation. Air sparging and application of nutrient solutions through venting piping was initiated after removal of the most mobile portion of LNAPL. Due to the rapid decrease of VOCs in the extracted air, SVE operations ended in 2000; venting in air injection mode was used

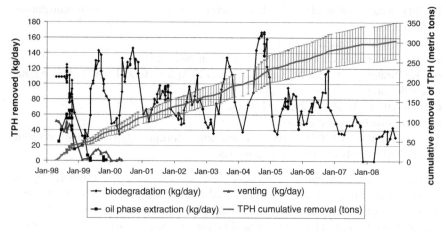

Fig. 21.4 Petroleum Hydrocarbons Removed from Field I Based on Remediation Mass Balance

afterwards with short operating periods in extraction mode for monitoring purposes [8]. Remediation of the unsaturated zone ended in 2001, when the remediation goals were reached. Since 2002, SVE has been used periodically for monitoring of respiratory gases and AS was used to stimulate in situ biological activity. Remedial progress was monitored during five soil sampling events, which took place in 1998, 2001, 2003, 2005 and 2008.

Figure 21.4 shows the removed amounts of petroleum hydrocarbons based on remediation monitoring results (remediation mass balance). A rapid decrease in the efficiency of physical remediation of the unsaturated zone is apparent on Fig. 21.4. During the period April 1998 – December 2008, free phase vacuum extraction removed 11.9 tons of LNAPL from the Field I, SVE removed 11.5 ton of VOCs and biodegradation removed 303.4 ton of petroleum hydrocarbons. The impact of seasonal fluctuation of temperatures on biodegradation intensity is also observable in Fig. 21.4. In 2001 and 2002, a decrease of intensity of biodegradation activity was observed, indicating substrate depletion. However, soil sampling in 2003 did not show a corresponding decrease in petroleum contamination levels. As a result of this decline, remedial efforts were intensified in 2003. The volume of air injected into the field by air sparging was increased from 100 to 250 m^3/h/ha.

This increase caused the radius of influence of the AS boreholes to increase [6] and conditions in previously insufficiently aerated parts of the field were optimized. Optimization resulted in higher biodegradation activity rates observed in 2003 and 2004 (Fig. 21.4). The biodegradation activity decrease in 2005 indicated substrate depletion in the remediation field. Soil sampling demonstrated that remediation goals in an area of 0.5 ha were achieved. Remediation ended in that sector and only 0.9 ha of the field remained in operation till finishing of the clean-up in 2008.

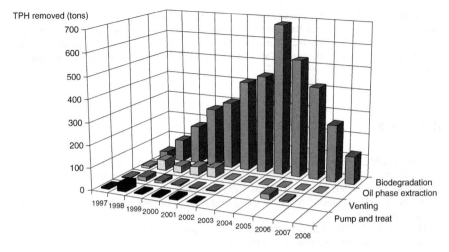

Fig. 21.5   Petroleum Hydrocarbons Removed from the Site by Different Technologies

21.2.2   Clean-Up Results at the Site

The clean-up technology was in operation on the site from 1997 till 2008, in 2009 and 2010 monitoring for verification of clean-up goals was held. Figure 21.5 shows clean-up efficiency of the clean–up technologies in each year of clean-up. Based on remediation mass balance, 4,046 metric tons of TPH (±15%), 93% by biodegradation, 4% by oil phase extraction (aka slurping), 2% by venting and 1% by pump-and-treat were removed from the site.

21.2.3   Reduction of Contamination Extent

Figures 21.1 and 21.2 shows changes in soil and groundwater contamination extent. Initial TPH soil mass estimation was 7,150 metric tons (±30%) in 1997, closure TPH soil mass estimation was 1,665 metric tons (±30%). Based on difference of these mass estimates, the amount removed from the site was 5,485 metric tons (±30%). The final soil and groundwater sampling for verification of reaching of clean-up targets was held in 2009 and 2010 and proved fulfilments of the goal.

21.3   Role of the Regulator in the Project

The whole clean-up project was funded by means of the Czech state budget. The work was performed by a private company managed by the Czech Ministry of Environment and supervised by the Czech Environmental Inspection. The final budget of the project was 420 million CZK (16.8 million euro).

Acknowledgements Scientific research was performed within the EU Seventh Framework Programme, Priority 6. Project Name BACSIN. Authors would like to thank the clean-up project managers Stanislava Proksova and Ferdinand Hercik for participation in soil data evaluation and soil sampling methods development and to the project team performing field measurements and soil sampling.

References

1. Anderson WC (1995) Vacuum vapour extraction, vol 542-B-94-002. US EPA
2. Boulding JR (1995) Practical handbook of soil, Vadose Zone and groundwater contamination assessment, prevention and remediation. CRC Press, Boca Raton
3. Havranek T (1993) Stastika pro biologicke a lekarske vedy Statistics for Biology and Medicine. Academia, Praha
4. Hercík F et al (2006) Etapová zpráva o sanaci ekologické zátěže na lokalitě Hradčany IX. etapa, " Periodical report on remediation at the Hradcany site, Period IX." EarthTech CZ Praha, 2006
5. Hercik F et al (2002) Etapova zprava sanace ekologicke zateze na lokalite Hradcany, VI. etapa, " Periodical report on remediation at the Hradcany site, Period VI."KAP, Praha, 2002
6. Johnson PC, Johnson RL, Bruce CL, Leeson A (2001) Advances in in situ air sparging/ biosparging. Bioremediation J 5(4):251–266
7. Landa I (2000) Aktualizace analyzi rizika lokalita Ralsko-letiste Hradcany, Risk Assessment of the Ralsko-airport Hradcany Site Ecoland, Praha, 2000
8. Leeson A, Hinchee R (1997) Soil bio venting Principles and Practice. Lewis Publishers, Boca Raton
9. Machackova J, Hercik F, Proksova S (2005) In situ bioremediation of jet fuel at the former Soviet Hradčany AFB, Proceedings of the 8th international in situ and on site bioremediation symposium, Baltimore USA, Battelle Memorial Institute, 2005
10. Machackova J, Wittlingerova Z, Vlk K, Zima J, Linka A (2008) Comparison of two methods for assessment of in situ jet-fuel remediation efficiency, water, air, & soil pollution, vol 187, Numbers 1–4/Jan, 2008, pp 181–194
11. Masak J, Machackova J, Siglova M, Cejkova A, Jirku V (2003) Capacity of the bioremediation technology for clean-up of soil and groundwater contaminated with petroleum hydrocarbons. J of Environ Sci and Health 38(10), Taylor & Francis 2003

Chapter 22
Health Significance of Safe Drinking Water

Mihail Kochubovski

Abstract The objective of the Drinking Water Directive (Council Directive 98/83/EC on the quality of water intended for human consumption) is to protect the health of the consumers in the European Union and to make sure the water is wholesome and clean. To make sure drinking water everywhere in the EU is healthy, clean and tasty, the Drinking Water Directive sets standards for the most common substances (so-called parameters) that can be found in drinking water. In the DWD a total of 48 microbiological and chemical parameters must be monitored and tested regularly. In principle WHO guidelines for drinking water are used as a basis for the standards in the Drinking Water Directive. Drinking-water quality management has been a key pillar of primary prevention for over one-and-a-half centuries and it continues to be the foundation for the prevention and control of waterborne diseases. Water is essential for life, but it can and does transmit disease in countries in all continents – from the poorest to the wealthiest. The most predominant waterborne disease, diarrhoea, has an estimated annual incidence of 4.6 billion episodes and causes 2.2 million deaths every year. The Protocol on Water and Health is the first major international legal approach for the prevention, control and reduction of water-related diseases in Europe. There have been some activities organized by the WHO and UNECE to tackle the issues about the influence of water pollution to the health and the environment. The aim of the paper is to review challenges, experiences, drinking water quality and health-related risks worldwide and particularly linked to small water supplies comparing with the public water supplying systems (up to 5,000 consumers) in the Republic of Macedonia. The difference with the public water supplying systems is protected catchments' area and organised maintenance of the water supplying network, as well continuous disinfection of drinking water. The final goal is to protect public health, especially to people living in rural areas.

M. Kochubovski (✉)
Institute of Public Health of the Former Yugoslav Republic of Macedonia,
St.50 Divizija No.6, 1000 Skopje, The Former Yugoslav Republic of Macedonia
e-mail: kocubov58@yahoo.com

F.F. Quercia and D. Vidojevic (eds.), *Clean Soil and Safe Water*, NATO Science for Peace and Security Series C: Environmental Security, DOI 10.1007/978-94-007-2240-8_22,
© Springer Science+Business Media B.V. 2012

22.1 Introduction

The objective of the Drinking Water Directive is to protect the health of the consumers in the European Union and to make sure the water is wholesome and clean. To make sure drinking water everywhere in the EU is healthy, clean and tasty, the DWD sets standards for the most common substances (so-called parameters) that can be found in drinking water. In the DWD a total of 48 microbiological and chemical parameters must be monitored and tested regularly. In principle WHO guidelines for drinking water are used as a basis for the standards in the Drinking Water Directive [2].

Drinking-water quality management has been a key pillar of primary prevention for over one-and-a-half centuries and it continues to be the foundation for the prevention and control of waterborne diseases. Water is essential for life, but it can and does transmit diseases in countries in all continents – from the poorest to the wealthiest. The most predominant waterborne disease, diarrhoea, has an estimated annual incidence of 4.6 billion episodes and causes 2.2 million deaths every year [16].

The Protocol on Water and Health is the first major international legal approach for the prevention, control and reduction of water-related diseases in Europe. There have been some activities organized by the WHO and UNECE to tackle the issues about the influence of water pollution to the health and the environment. One in ten citizens of the EU receives drinking-water from small or very small systems, including private wells. Today's national and international policy frameworks recognize that further attention to this topic is needed. The Protocol on Water and Health in article 5 requires that: "Equitable access to water, adequate in terms both of quantity and of quality, should be provided for all members of the population, especially those who suffer a disadvantage or social exclusion" [14].

At present, nearly 140 million people – 16% of Europe's population – still live in homes that are not connected to a drinking-water supply and about 85 million people do not have access to adequate sanitation. According to WHO, more than 13,000 children under the age of 14 die every year from water-related diarrhoea, mostly in Eastern Europe and Central Asia. Access to safe drinking water and to improved sanitation remains a challenge in several countries of the UNECE region, with some of them actually regressing instead of progressing towards the MDG (Millennium Development Goal). Mortality and morbidity related to unsafe water and inadequate sanitation remain unacceptably high; and water resources are often used in an inefficient manner.

Vulnerable population groups may be particularly susceptible to water-related hazards. Outbreaks have been associated with both microbial and chemical contamination. A significant proportion of such waterborne disease is associated with contamination within buildings. This can arise from:

- direct contamination through faults in water systems (e.g. bird and small animal droppings into storage tanks) or leaching from inappropriate materials or corrosion (e.g. copper, lead, nickel, cadmium);
- indirect contamination through cross-connections between drinking-water systems and contaminated water or chemical storages;

- growth of indigenous microbes (e.g. Pseudomonas aeruginosa, non-tuberculous Mycobacteria and legionellae)

WHO has identified that the benefits of all interventions to reduce risks from unsafe water outweigh costs by substantial margins [8].

Once potential hazards and hazardous events have been identified, the levels of risk need to be assessed so that priorities for risk management can be established. Risk assessments need to consider the likelihood and severity of hazards and hazardous events in the context of exposure (type, extent and frequency) and the vulnerability of those exposed. Although many hazards may threaten water quality, not all will represent a high risk. The aim should be to distinguish between high and low risks so that attention can be focused on mitigating risks that are more likely to cause harm. Ingress of microbial and chemical contaminants can be caused by a range of hazardous events, including contamination of water sources by human and animal waste, industrial spills and discharges, inadequate treatment, inappropriate storage, pipe breakages and accidental cross-connections. In well-managed systems, bio-films will be thin and relatively well contained. Concern arises when these bio-films become too thick and start to disseminate throughout the system. Organisms in established bio-films can be difficult to remove. Poorly managed building water systems are prone to colonization, and bio-films can develop within pipes and on components such as washers, thermostatic mixing valves and outlets. Bio-films are extremely difficult to remove from all parts of the system once they are established, and they can be resistant to disinfectants, such as chlorine. Well-managed disinfection regimes that maintain disinfectant residuals through water systems can inactivate potential pathogens released into the aqueous phase, but this protection is lost if disinfectant residuals fall below effective levels. Unsuitable materials and equipment used in water systems may release hazardous substances into drinking-water. The chemicals could be contaminants in the materials, be leached during initial use, or be leached due to elevated corrosion. Stagnation of water within the building system can increase concentrations of hazardous chemicals released from materials. Intermittent use of end-of-plumbing fixtures (e.g. drinking-water coolers in schools) can result in the presence of elevated concentrations of heavy metals such as copper from copper piping or lead from brass fixtures [3].

The World Health Organization (WHO) Guidelines for drinking-water quality [18] define the reduction of disease and outbreaks as health outcome targets. Reducing disease provides the most direct evidence of the success of WSPs, while continued disease provides evidence that WSPs are inadequate and require modification. While the immediate response to detection of disease is necessarily reactive, the subsequent responses can be proactive in identifying and eliminating building-specific and systemic risks. Many countries have mechanisms for surveillance and reporting of communicable diseases. The importance of these mechanisms is reinforced by the International Health Regulations, which call for Member States to apply and – where necessary – strengthen capabilities for surveillance, reporting, notification and communication of infectious disease.

While surveillance programmes often include waterborne organisms, specific surveillance of water as a source of disease is generally not well developed or coordinated. Improvement in access to safe water and adequate sanitation, along with promotion of good hygiene practices (particularly hand washing with soap), can help prevent childhood diarrhoea. In fact, an estimated 88% of diarrhoeal deaths worldwide are attributable to unsafe water, inadequate sanitation and poor hygiene [1].

Water, sanitation and hygiene programmes typically include a number of interventions that work to reduce the number of diarrhoea cases. These include: disposing of human excreta in a sanitary manner, washing hands with soap, increasing access to safe water, improving water quality at the source, and treating household water and storing it safely [15].

The use of improved sources of drinking-water is high globally, with 87% of the world population and 84% of the people in developing regions getting their drinking-water from such sources. Even so, 884 million people in the world still do not get their drinking-water from improved sources, almost all of them in developing regions. Sub-Saharan Africa accounts for over a third of that number, and is lagging behind in progress towards the MDG target, with only 60% of the population using improved sources of drinking-water despite an increase of 11 percentage points since 1990. Regional Priority Goal I is to prevent and significantly reduce the morbidity and mortality arising from gastrointestinal disorders and other health effects, by ensuring that adequate measures are taken to improve access to safe and affordable water and sanitation for all children. Population access to improved water sources, sanitation and wastewater treatment has increased over the past two decades in most Member States. Progress in many countries in the east of the Region is, however slow, giving rise to important health inequalities. Water-related diseases remain a burden for people throughout the Region, including in the most economically developed countries. To reduce these diseases, a change is required from the present system of controlling drinking-water solely at the tap towards quality management along the production and distribution continuum for capture to tap. Thus there needs to be a shift in policy approach from penalties to active support. Gaps remain in our understanding of the distribution and cause of water-related diseases. Harmonized surveillance systems for waterborne diseases and outbreaks are needed throughout the Region, as are systems for monitoring health risks related to bathing water. It is particularly important to maintain a core of expertise to advice on and conduct outbreak investigation; testing, implementing and revising procedures in cooperation with other actors; and updating regulations and policy. Legislation adopted in the framework of the EU acquis communautaire is an important policy driver throughout the Region. The UNECE/WHO Protocol on Water and Health offers the Region-wide legal framework for the reduction of water-related diseases, integrated water resource management, a sustainable water supply compliant with WHO's Guidelines for drinking water quality and adequate sanitation for all. Climate change is adding to the challenge of providing sustainable water and sanitation services. Urgent action is required to assess systematically the climate change resilience of water supply and sanitation utilities, and to include the effects of climate change in water safety plans [16].

22.2   Material and Methods

It has been provided an evaluation of international references on drinking water and sanitation and related health impacts, as well a community survey on the whole country with a goal to review water safety in different settings (urban and rural areas).

22.3   Aim

The aim of the paper is to review challenges, experiences, drinking water quality and health-related risks worldwide and particularly linked to small water supplies comparing with the public water supplying systems (up to 5,000 consumers) in the Former Yugoslav Republic of Macedonia.

The difference with the public water supplying systems is protected catchments' area and organized maintenance of the water supplying network, as well continuous disinfection of drinking water.

The goal is to protect public health, especially to people living in rural areas.

22.4   Results and Discussion

In 2009 1,545,655 inhabitants or 66.8% of total population in the Former Yugoslav Republic of Macedonia is supplied with drinking water by central water supplying systems managed by Communal Public Enterprises that fulfil the legal obligations related to providing and drinking water safety control.

482,952 inhabitants in rural areas (52.3% of rural population/22.16% of total population in the country) are supplied from local public water supplying systems, with which there is no always management by Communal Public Enterprise.

139,522 inhabitants (15.1% of rural population/6.4% of total population in the country) are supplied from local water supplying objects (springs, Norton's pumps, drilled wells etc.).

There were performed 1,038 sanitary-hygienic inspections on public water supplying objects in 2009, and it was found 18.67% improper physico-chemical water samples, as well 32.69% microbiological water samples.

- In 2009 there was a significant difference between public water supplying systems in the cities with 1.18% improper microbiological water samples, and 9.17% in the villages connected to the water supplying systems from the cities
- But the highest difference was realized in the drinking water samples from the rural areas with their own water supply systems, 29.6% drinking water samples have been microbiologically improper, and especially in the villages with private water supplying objects – 42.63% drinking water samples have been microbiologically improper [12].

In accordance to the Law on drinking water supplying and disposal of urban waste water it is regulated the process of distribution of drinking water to the consumers, as well the disposal of urban waste water. Communal Public Enterprises are responsible for ensuring the quality and quantity of drinking water, as well for disposal and treatment of urban waste water [9].

But, there is a lack of enforcement for communities in rural areas, without Communal Public Enterprises dealing with these issues, and very often drinking water supplying is not managed, or there is not regular disinfection of drinking water. Small water supply object are at the utmost public health risk from bacteriological pollution, and physico-chemical contamination. Despite the fact that public health authorities are requesting regular disinfection of drinking water, some Local Self-Government do not priority to solve this issue. The Institute of Public Health with its own network of Public Health Institutes has been working on environmental health awareness raise with stakeholders from the communities during whole period, especially in 2009–2010. Recently have been introduced counter measures by Communal Public Enterprises with a goal to raise people's awareness about the need to spare drinking water, especially in the eastern part of the country during summer season. Also, there are lot of efforts for preparation and secure of the catchments' area surrounding drinking water sources and water supplying objects. The Institute of Public Health is dealing with this topic, as well the monitoring of drinking water quality. There was a National Action Programme for Improvement of sanitary-hygienic situation in rural areas in the Former Yugoslav Republic of Macedonia since 1971–1991. The leader of that was the Republic Institute for Health Protection-Skopje, financed by Water Economy Secretariat and Health Insurance Fund. During this Action Programme have been built water supplying networks in 850 villages, as well 25 sewerages. Since 1991–2006 have been built new water supplying networks in 90 villages. In 1971 access to safe drinking water in the Former Yugoslav Republic of Macedonia was 64%, and after 1971–1991 National Action Programme and efforts from 1991–2003 access to safe drinking water in 2003 has increased to 93%. Institute of Public Health of the Former Yugoslav Republic of Macedonia and its laboratories have been accredited for ISO 17025. As well the regional Public Health Centres (10) are conducting the accreditation for ISO 17025 but, for basic methods of food quality investigation. The surveillance system is aimed at prevention and early alert, as well outbreak detection and control/assessment of contagious diseases. There is already established an ALERT System supported by WHO in 2005. IPHRM and 10 Regional PHC are responsible for monitoring of drinking-water quality. They report to the Food Directorate, part of the Ministry of Health. FD is established and started to work in 2005 [10].

On 28 July 2010, the UN General Assembly adopted a resolution declaring that access to clean water and sanitation is a human right. Less than month ago, the UN Human Rights Council has – by consensus – adopted a decision affirming this right. The Protocol on Water and Health embodies the close linkages between human rights, health, environmental protection and sustainable development. The Protocol guides its Parties on how to translate the human right to water into practice, and how safe, acceptable, affordable, accessible and sufficient water can be provided, in accordance with the principles of non-discrimination and of transparency of

information. The second session of the Meeting of the Parties to the UNECE/
WHO-EURO Protocol on Water and Health to the Convention on the Protection and
Use of Transboundary Watercourses and International Lakes has taken place on
23–25 November 2010 in Bucharest, Romania. Parties have discussed future activities to promote the exchange of experience on measures supporting equitable access
to water and thus addressing affordability issues and access in remote communities
[7]. There is much evidence that in rural areas with mainly small scale water supply
exist problem with organic pollution of drinking water. The outcome of the above
mentioned causes is bacteriological pollution of drinking water, as well increase of
the COD (chemical oxygen demand) and other physico-chemical parameters related
to pollution with nutrients. The most effective means of consistently ensuring the
safety of a drinking-water supply is through the use of a comprehensive risk assessment and risk management approach that encompasses all steps in water supply
from catchment to consumer. Such approaches are called Water Safety Plans [14].

OFDA/CRED International Disaster Database for the period 1990–2009 present
the effects of extreme weather events and disasters such as floods and droughts in
the European region, as number of events – 413 (floods) and 36 (droughts), death
– 3,912 (floods) and 2 (droughts), affected population – 12,137,319 (floods) and
15,875,965 (droughts), and economic damage in thousands of USD – 84,072,159
(floods) and 15,082,309 (droughts) [4].

In 2006, nearly 140 million people (16%) in the Eastern Europe, Caucasus and
central Asia (EECCA) region still did not have a household connection to a drinking-
water supply, over 41 million people (5%) did not have access to a safe drinking-
water supply and 85 million people (10%) did not have improved sanitation. In the
Pan-European Region, approximately 30% of the total population live in rural areas,
in which small-scale water supplies prevail and where access to improved drinking-
water sources varies between 61% and 100% [15].

The United Nations Millennium Development Goal (MDG) target 7c calls on
countries to halve, by 2015, the proportion of people without sustainable access to
safe drinking water and basic sanitation. The Joint Monitoring Programme (JMP)
for Water Supply and Sanitation of WHO and the United Nation Children's Fund
(UNICEF) is mandated to monitor progress towards that MDG. In the pan-European region, approximately 30% of the total populations live in rural areas, in which
small-scale water supplies predominantly prevail. Access to improved drinking-
water sources in countries of this region varies between 70% and 100% of the total
population, and in rural areas between 61% and 100%. Of the population in urban
areas, 1% is without access to improved drinking-water sources; however, in rural
areas, this is the case for 6% of the population, or approximately 14 million people.
Some countries – such as Wales and England in the United Kingdom, as well as
Hungary – have revised their drinking-water regulations to require water suppliers
to implement WSPs; in Switzerland, Iceland, Sweden and the Former Yugoslav
Republic of Macedonia, drinking water is classified as a foodstuff, and water suppliers are required to comply with food hygiene regulations and to prepare safety
plans on the basis of HACCP principles; and the European Commission is considering including WSP requirements in the forthcoming revision of the EU Drinking
Water Directive [17].

The new 2009 EU water initiative (EUWI) Annual Report shows that European institutions have doubled their support to the sector since 2002 and that there is a changing trend from assisting large water supply and sanitation systems to basic systems (from 22% of the aid in 2007 to 37% in 2009), signalling an increased focus on the most in need of water and sanitation. The average amount spent in the sector by all EU donors is over one billion Euros annually [5].

22.5   Conclusion and Recommendations

The health risk to the local population is much higher in those rural areas without a public water supplying systems, due to lack of continuous disinfection of drinking water, as well lack of proper management, maintenance and protection of the catchments' area.

That is why the environmental health professionals in cooperation with the local stakeholders shall put a lot of efforts to minimize and prevent the potential risk from microbiological and physico-chemical pollution of drinking water and possible adverse health outcomes, especially water-borne diseases.

In order to minimize possible adverse health impact to people living in rural areas that receives drinking-water from small or very small water-supplying systems and/or without a public water supplying systems each country should:

- Promote best practice to manage drinking water quality effectively;
- Provide continuous drinking water quality monitoring, not only in urban and rural areas with public water supplying systems, but also in rural communities with small-scale water supplies;
- Highlight critical needs and key arguments for taking action to improve the situation for small scale water supplies;
- Implement the improved surveillance of water related diseases;
- Promote the role of the National Policy Dialogue on Integrated Water Resources Management in setting targets under the UNECE&WHO Protocol on water and Health;
- Strengthen the evidence based medicine for a changing climate and its potential impact to water, health and diseases

UN on World Water Day 22nd March 2011 Water for cities: responding to the urban challenges has stressed the following issues:

- Investments in infrastructure have not kept up with the rate of urbanization, while water and waste services show significant underinvestment. The central problem is therefore the management of urban water and waste. Piped water coverage is declining in many settings, and the poor people get the worst services, yet paying the highest water prices.
- Few urban authorities in developing countries have found a sustainable solution to urban sanitation, and utilities cannot afford to extend sewers to the slums, nor can they treat the volume of sewage already collected.

- 828 million people live in slums or informal settlements that are scattered around the world's cities; the biggest challenge is to provide these people with adequate water and sanitation facilities.
- The urban poor pay up to 50 times more for a litre of water than their richer neighbours, since they often have to buy their water from private vendors.

References

1. Black RE, Morris S, Bryce J (2003) Where and why are 10 million children dying every year? The Lancet 361(9376):2226–2234
2. Council Directive (1998) 98/83/EC on the quality of water intended for human consumption, Brussels
3. Cunliffe D, Bartram J, Briand E, et al. (2011) Water safety in buildings. World Health Organization, Geneva, pp 1–94
4. EM-DAT: The OFDA/CRED International disaster database, www.em-dat.net – Université catholique de Louvain, Brussels, © 2009 CRED
5. http://www.euwi.net/files/Press_release_final2_con_imagine.pdf
6. http://www.unece.org/env/water/text/text_protocol.htm
7. http://www.unece.org/env/water/whmop2.htm
8. Hutton G, Haller L (2004) Evaluation of the costs and benefits of water and sanitation improvements at the global level. World Health Organization, Geneva (http://www.who.int/water_sanitation_health/wsh0404.pdf)
9. Law on drinking water supplying and disposal of urban waste water (2004) Official Gazette of RM No.68/2004
10. Mihail Kochubovski (2007) Implementation of the protocol on water and health in Macedonia. A handbook for teachers, researchers, health professionals and decision makers "Health promotion and disease prevention". Hans Jacobs Publishing Company, Lage, pp 71–85. ISBN 978-3-89918-169-2
11. Progress and challenges on water and health: the role of the Protocol on Water and Health (2009) World Health Organization, Copenhagen, p 3
12. Report of preventive health care in the republic of Macedonia (2010) Institute of Public Health of the Republic of Macedonia
13. Small-scale water supplies in the pan-European region, background-challenges-improvements (2010) World Health Organization, Copenhagen, pp 8–43
14. UNECE (2010) Meeting of the parties to discuss progress regarding water and health in the Pan-European region, Issue No. 398, 8–12 Nov 2010
15. UNICEF, WHO (2009) Diarrhoea: why children are still dying and what can be done, vol 11. WHO Press, Geneva
16. WHO (2010) Water for health WHO Guidelines for drinking water quality, Geneva
17. World Health Organization (2010) Health and environment in Europe: progress assessment, vol 5. WHO Regional Office for Europe, Copenhagen
18. World Health Organization (WHO) (2008) Guidelines for drinking-water quality, Geneva